# Standard Mechanical and Electrical Details

# Standard Mechanical and Electrical Details

Jerome F. Mueller

**McGraw-Hill Book Company**

New York  St. Louis  San Francisco  Auckland
Bogotá  Hamburg  Johannesburg  London
Madrid  Mexico  Montreal  New Delhi  Panama
Paris  São Paulo  Singapore  Sydney  Tokyo  Toronto

**Library of Congress Cataloging in Publication Data**
Mueller, Jerome F
    Standard mechanical and electrical details.

    Includes index.
    1. Buildings—Mechanical equipment—Handbooks,
manuals, etc.   2. Buildings—Electric equipment—Hand-
books, manuals, etc.   I. Title.
TH6010.M83        696         80-11161
ISBN   0-07-043960-5

1234567890     HD!HD     8987654321

The editors for this book were Tyler G. Hicks and Geraldine Fahey, the
designer was Mark E. Safran, and the production supervisor was Thomas
G. Kowalczyk. It was set in Optima by Bi-Comp, Incorporated.

Printed and bound by Halliday Lithograph.

# Contents

# Preface

As every consultant knows, there is no true standard detail. Each piece of equipment needs something to isolate it from the system that it serves. Beyond that, the elaboration of isolating devices has a very wide range of alternatives.

Secondly, there is no true standard method of making connections. The standard is what you and I and the applicable codes say is the standard. Generally, this means, beyond the code limitation, the standard is what we think properly fits a given situation. Another factor may be what our client insists he wants provided.

If this book truly covered all possible cases, it could easily be ten times as big as it is. Since it does not cover all possible cases, the obvious question is, why not? In my judgment, the answer is that the book tries to reach a common denominator of the most frequently used details.

Since the book, in the author's own words, is neither complete nor standard, the final question is obviously, of what value is it? The answer is twofold. If you already have these details in a neat, orderly, usable, and readily accessible set of files, you have no need for this book.

On the other hand, if your office is like my office, then the words are more likely you "had" the detail, but can't find it. After several hours or more are wasted looking for it, you curse under your breath, take another half hour or more, and reinvent the detail one more time.

The details presented are based on the safe assumption that you, the user, are neither lazy nor stupid. Quite the contrary. You are working very hard against difficult deadlines and are very much above the average in intelligence. You don't need a simple, canned solution to something you already know. What you'd like to have is a quick frame of reference, a good guide to something you know but can't readily recall.

In the book are details that I seriously debated including. Not in this book are details that I perhaps should have included. Between those two limits are the details that my 29 years of experience tell me are probably the most frequently or infrequently used. In either case, if your practice covers the mechanical and electrical phases of building design, I think you'll be glad I included the ones I did.

Are these details "the" way to do the connection illustrated? Absolutely not! They are a way of making the connection. Can you improve on the detail? Depending upon your circumstances and design criteria you can, you should, and I am sure you will. I am not your teacher. I'm one of you who also struggles to keep the firm going.

So if you will accept these details as a clue sheet rather than a master list, I think that regardless of your position you will be able to combine my ideas with yours to much more quickly, efficiently, and economically achieve the result you want.

Finally, you are going to need more than one copy of this book. In the case of my firm, I plan on seven copies to be exact. Two are going to be in three separate departments, and one I am going to carefully put away. When employees leave, books and catalogs also seem to sometimes take their leave. Since I'm tired of inventing the details, I'm going to make certain that I always have my copy where I know it can be found—in my desk!

Jerome F. Mueller

vii

# 1. HVAC Systems

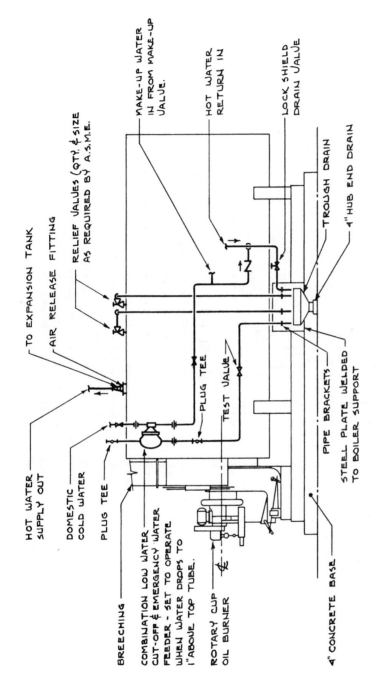

**FIGURE 1-1**

INSTALLATION of the TWO PASS, OIL FIRED, SCOTCH TYPE, HOT WATER BOILER

Figure 1-1 shows a combination low water cutoff and emergency water feeder for a hot water boiler. The combination low water cutoff and water feeder should be used to cut off the burner and supply make-up water under full domestic water pressure when the water in the boiler falls to 2" above the top tubes. The water feeding feature of this device should not be considered as a make-up water feeder which normally comes in through a pressure-reducing valve to maintain a predetermined pressure in the system, but rather as an emergency feed.

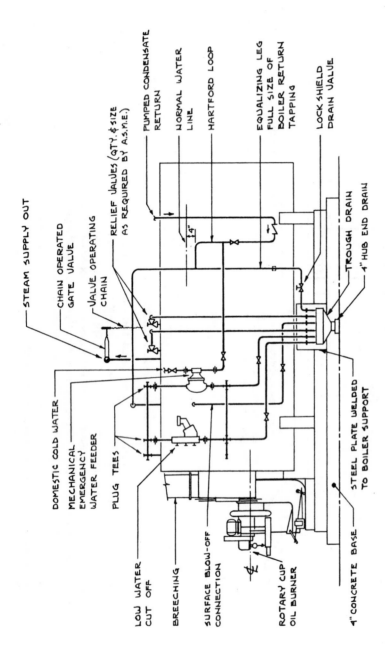

STEAM SUPPLY OUT

CHAIN OPERATED GATE VALVE

VALVE OPERATING CHAIN

RELIEF VALVES (QTY. & SIZE AS REQUIRED BY A.S.M.E.)

PUMPED CONDENSATE RETURN

NORMAL WATER LINE

HARTFORD LOOP

EQUALIZING LEG FULL SIZE OF BOILER RETURN TAPPING

LOCK SHIELD DRAIN VALVE

TROUGH DRAIN

4" HUB END DRAIN

DOMESTIC COLD WATER

MECHANICAL EMERGENCY WATER FEEDER

PLUG TEES

LOW WATER CUT OFF

BREECHING

SURFACE BLOW-OFF CONNECTION

ROTARY CUP OIL BURNER

STEEL PLATE WELDED TO BOILER SUPPORT

4" CONCRETE BASE

## FIGURE 1-2

INSTALLATION of the TWO PASS, OIL FIRED, SCOTCH TYPE, STEAM BOILER

Figure 1-2 shows the standard trim on a steam fired boiler. The boiler should always be equipped with the following minimum trim: a high limit pressure stat, an operating pressure stat, a pressure gauge, a pump starter, low water cutoff and alarm, emergency water feeder, and steam relief valves of the number and capacity as required by the American Society of Mechanical Engineers Code. Boiler stop valves are required to be of the OS&Y rising stem type and make-up water feed must be connected on the boiler side of all valves.

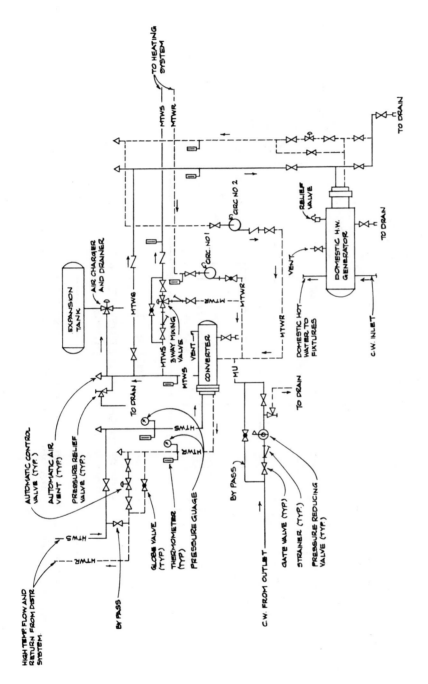

**FIGURE 1-3**

TYPICAL CONVERTER CONNECTION FOR HTW SYSTEM

Figure 1-3 shows typical converter connections utilizing high temperature hot water reduced to a lower temperature for domestic or other use and in addition to provide domestic water for plumbing fixture use. Not all of the valves are noted where their use is felt to be self-evident. Note that bypass valves around control valves are globe valves.

9

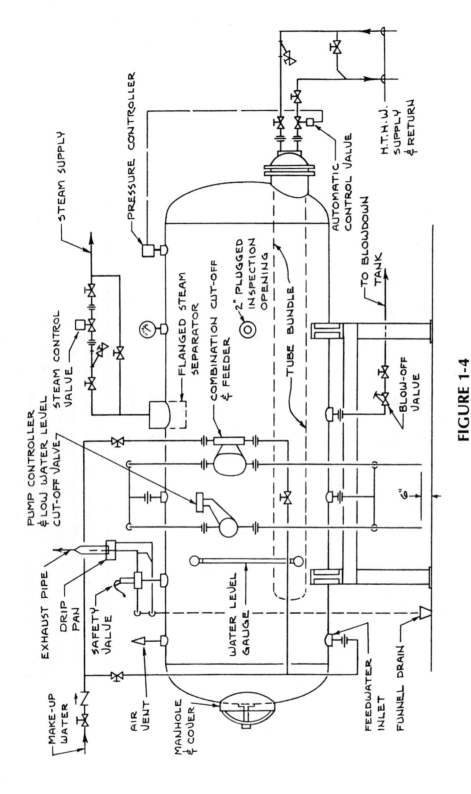

**FIGURE 1-4**

TYPICAL HIGH TEMPERATURE HOT WATER TO
STEAM CONVERTOR PIPING DETAIL

Figure 1-4 is frequently used in systems where the basic distribution is
high temperature hot water, and is useful where there are steam requirements. Size the safety valves per code.

10

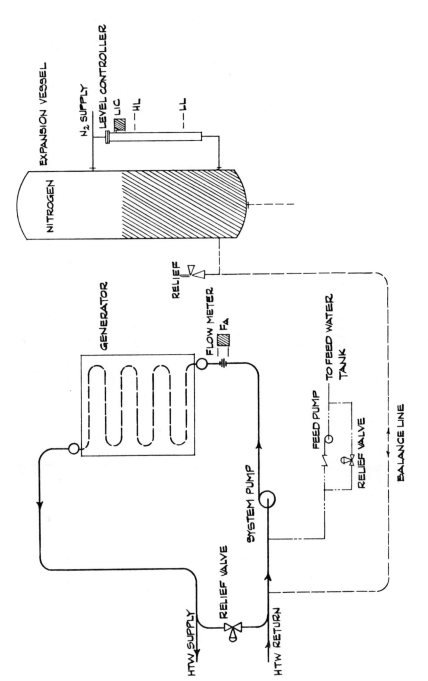

**FIGURE 1-5**

HIGH TEMPERATURE WATER CYCLE INERT · GAS
PRESSURIZATION SINGLE PUMP

Figure 1-5 is not intended to show the exact connection details of system components, but rather to emphasize the relationship of one to the other. It is particularly important that the engineer carefully calculate the required expansion space in the nitrogen pressurized expansion tank.

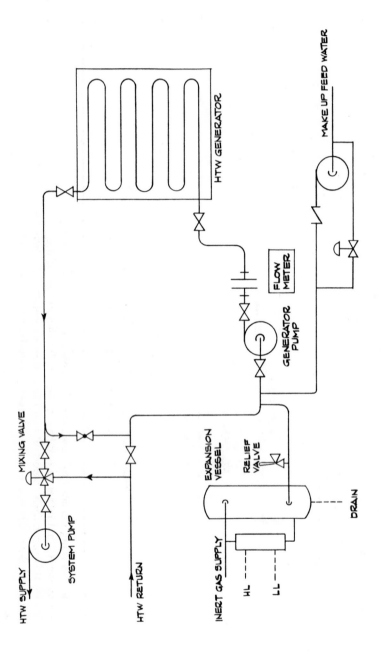

**FIGURE 1-6**

HIGH TEMPERATURE WATER CYCLE - INERT - GAS PRESSURIZED
TWO PUMP

Figure 1-6 is similar in general approach to Figure 1-5. Again, the intent is
not to show every connecting detail, but to make the designer aware of
the general arrangement of the items in the system shown.

12

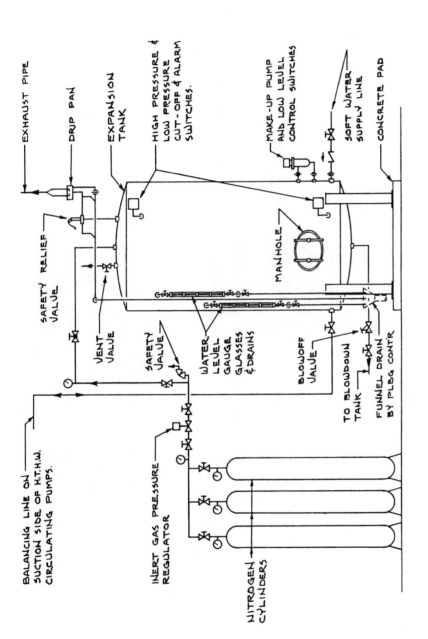

**FIGURE 1-7**

TYPICAL HIGH TEMPERATURE HOT WATER
EXPANSION TANK PIPING

Figure 1-7 is a more specific detail of the expansion tank noted in Figures
1-5 and 1-6 and the details of connections are much more explicit in their
relationship to each other. All components must be carefully sized to fit
the particular application.

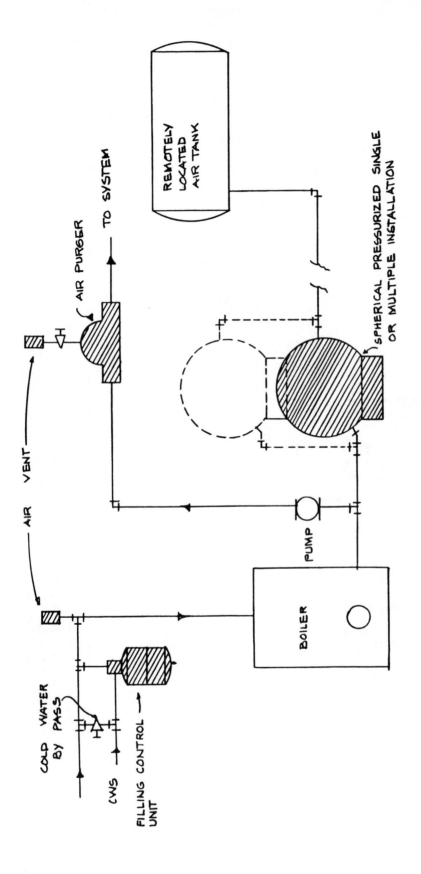

**FIGURE 1-8**

FACTORY PRESSURIZED SYSTEM
WITH REMOTE AIR TANK

Figures 1-8, 1-9, and 1-10 show the use of factory pressurized expansion tanks in lieu of the standard expansion tank which uses the system pressure to compress air. Various arrangements are shown for boilers, chillers, and combinations of boilers and chillers.

14

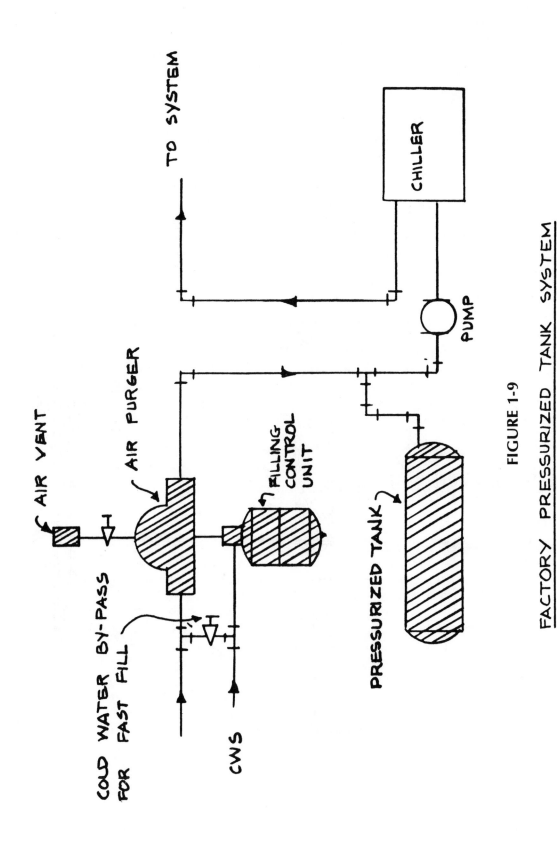

TO SYSTEM

CHILLER

PUMP

AIR VENT

AIR PURGER

FILLING CONTROL UNIT

PRESSURIZED TANK

COLD WATER BY-PASS FOR FAST FILL

CWS

FIGURE 1-9

FACTORY PRESSURIZED TANK SYSTEM

FOR CHILLED WATER SUPPLY

15

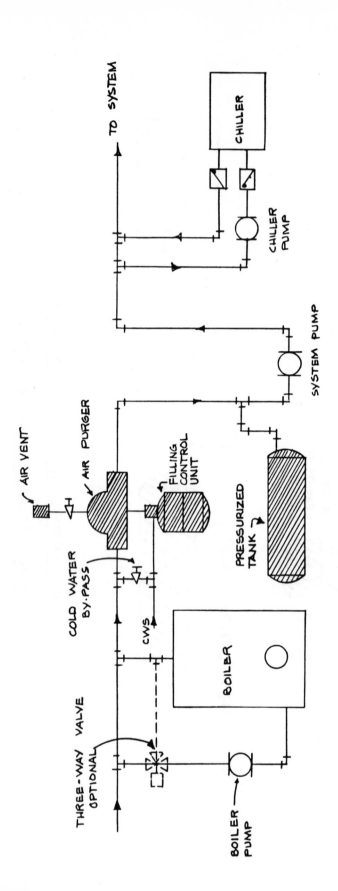

**FIGURE 1-10**

FACTORY PRESSURIZED TANK INSTALLATION
WITH DUAL WATER TEMPERATURES & SEPARATE BOILER
& CHILLER PUMPS

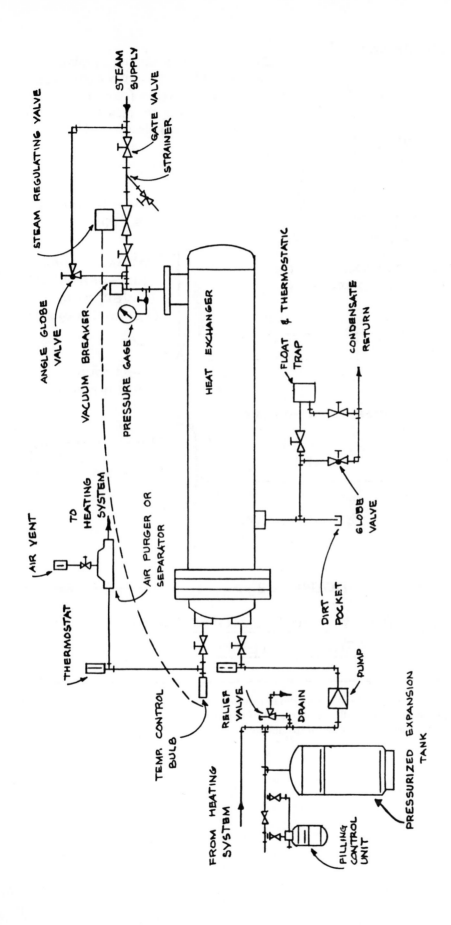

**FIGURE 1-11**

CONVERTOR/FACTORY PRESSURIZED TANK INSTALLATION

Figure 1-11 is a little more complex where, again, the factory pressurized expansion tank is used, but this time in a heat exchanger type hot water system in lieu of the standard boiler arrangement.

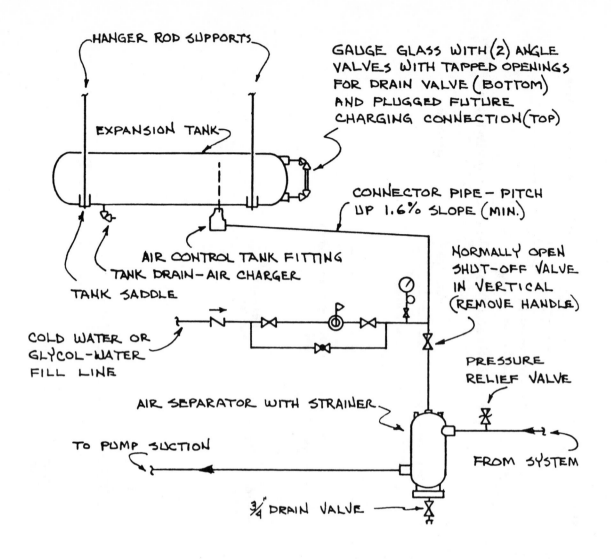

**FIGURE 1-12**

AIR CONTROL & PIPING CONNECTIONS
FOR WATER SYSTEMS

Figure 1-12 is the standard air pressurized tank. This detail can be used
for either a water filled system or a glycol filled system. All of the connections and fittings are fairly standard for this type of installation.

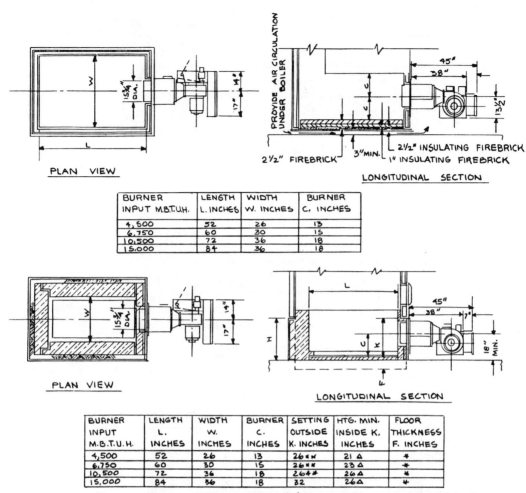

PLAN VIEW

LONGITUDINAL SECTION

2½" FIREBRICK   3" MIN.   2½" INSULATING FIREBRICK
                          1" INSULATING FIREBRICK

| BURNER INPUT M.B.T.U.H. | LENGTH L. INCHES | WIDTH W. INCHES | BURNER C. INCHES |
|---|---|---|---|
| 4,500 | 52 | 26 | 13 |
| 6,750 | 60 | 30 | 15 |
| 10,500 | 72 | 36 | 18 |
| 15,000 | 84 | 36 | 18 |

PLAN VIEW

LONGITUDINAL SECTION

| BURNER INPUT M.B.T.U.H. | LENGTH L. INCHES | WIDTH W. INCHES | BURNER C. INCHES | SETTING OUTSIDE K. INCHES | HTG. MIN. INSIDE K. INCHES | FLOOR THICKNESS F. INCHES |
|---|---|---|---|---|---|---|
| 4,500 | 52 | 26 | 13 | 26 ** | 21 △ | * |
| 6,750 | 60 | 30 | 15 | 26 ** | 23 △ | * |
| 10,500 | 72 | 36 | 18 | 26 ** | 26 △ | * |
| 15,000 | 84 | 36 | 18 | 32 | 26 △ | * |

NOTE: * CHAMBER FLOOR THICKNESS (F.) REQUIRED WILL BE DETERMINED BY FIRING RATE, FIRING CYCLE, MATERIAL USED, & PROTECTION REQUIREMENTS OF BOILER ROOM FLOOR AND BOILER FOOTINGS.

** MINIMUM FOR BURNER-FLOOR CLEARANCE OUTSIDE SETTING
△ MINIMUM WATERLEG TO CHAMBER FLOOR.

## FIGURE 1-13

## BURNER REFRACTORY FURNACE DIMENSIONS

Figures 1-13 through 1-18 are typical firebrick setting drawings for the standard average smaller boiler installation. These details are to be used as needed in a given situation and they illustrate certain basic sizing situations. For a small cast-iron boiler, Figure 1-13 gives typical sizing and arrangements. The ventilated floor, which is common to most small to medium sized boilers, is shown in Figures 1-14 and 1-15. In addition, Figure 1-16 shows a typical ventilated floor and continues with elevation drawings where an existing boiler of insufficient combustion space is changed to an oil fired situation and where the boiler base has to be either proved in a pit or the boiler raised to suit the conditions of combustion space. This is covered on Figures 1-16, 1-17, and 1-18. The actual size of the chamber is a design problem. Generally, the width and height are approximately equal and the length is some 3 times the width.

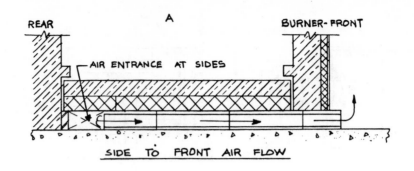

SIDE TO FRONT AIR FLOW

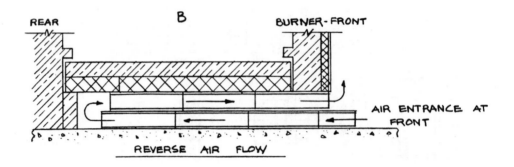

REVERSE AIR FLOW

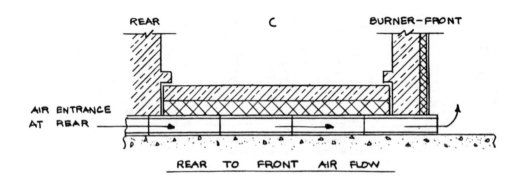

REAR TO FRONT AIR FLOW

**FIGURE 1-14**

VENTILATED FLOOR AIR FLOW

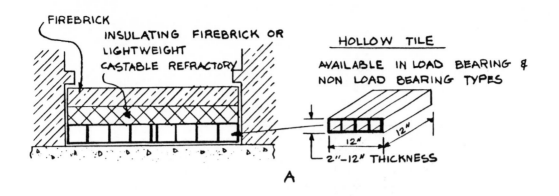

FIREBRICK
INSULATING FIREBRICK OR LIGHTWEIGHT CASTABLE REFRACTORY

HOLLOW TILE

AVAILABLE IN LOAD BEARING & NON LOAD BEARING TYPES

12"
12"
2"-12" THICKNESS

A

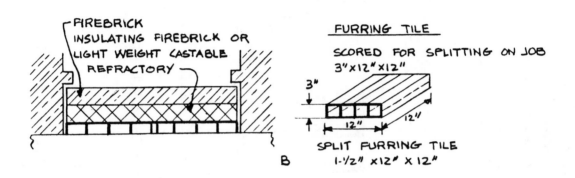

FIREBRICK
INSULATING FIREBRICK OR LIGHT WEIGHT CASTABLE REFRACTORY

FURRING TILE

SCORED FOR SPLITTING ON JOB
3"x12"x12"

3"
12"
12"

SPLIT FURRING TILE
1-1/2" x 12" x 12"

B

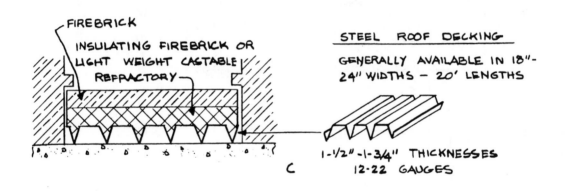

FIREBRICK
INSULATING FIREBRICK OR LIGHT WEIGHT CASTABLE REFRACTORY

STEEL ROOF DECKING

GENERALLY AVAILABLE IN 18"-24" WIDTHS — 20' LENGTHS

1-1/2"-1-3/4" THICKNESSES
12-22 GAUGES

C

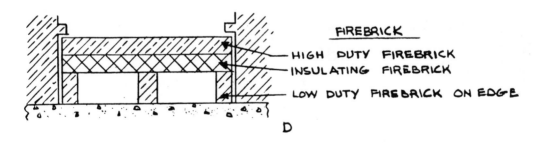

FIREBRICK

HIGH DUTY FIREBRICK
INSULATING FIREBRICK
LOW DUTY FIREBRICK ON EDGE

D

**FIGURE 1-15**

VENTILATED FLOOR SUPPORTS

21

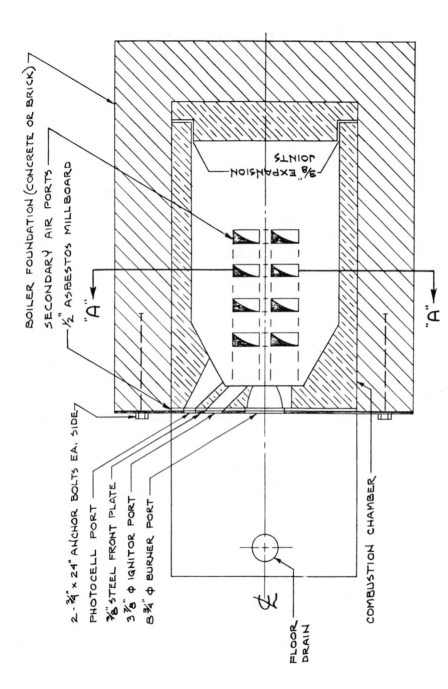

PLAN AT BURNER CENTER LINE

**FIGURE 1-16**

TYPICAL OIL FIRED COMBUSTION CHAMBER on EXISTING BOILER

SHEET 1 of 3

22

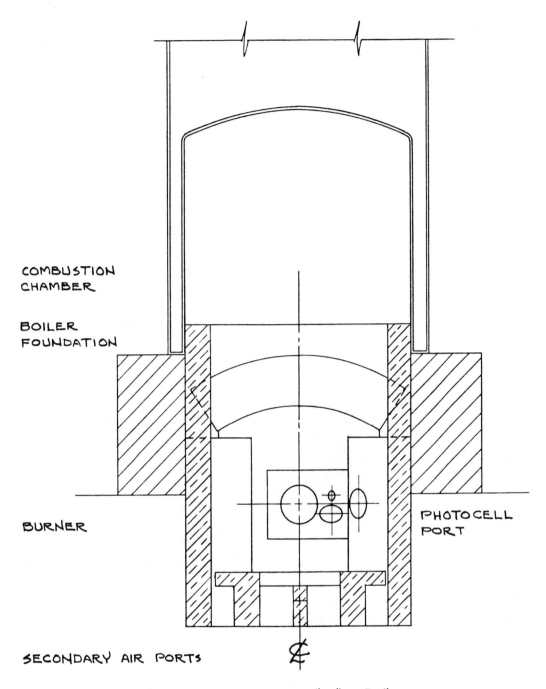

COMBUSTION
CHAMBER

BOILER
FOUNDATION

BURNER

SECONDARY AIR PORTS

PHOTOCELL
PORT

SECTION AT "A"-"A"

**FIGURE 1-17**

TYPICAL OIL FIRED COMBUSTION CHAMBER
ON EXISTING BOILER

SHEET 2 OF 3

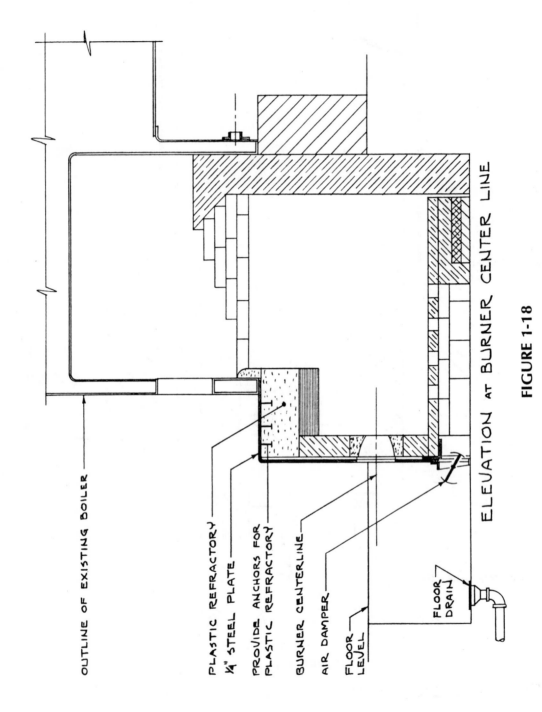

OUTLINE OF EXISTING BOILER

PLASTIC REFRACTORY

¼" STEEL PLATE

PROVIDE ANCHORS FOR PLASTIC REFRACTORY

BURNER CENTERLINE

AIR DAMPER

FLOOR LEVEL

FLOOR DRAIN

ELEVATION at BURNER CENTER LINE

**FIGURE 1-18**

TYPICAL OIL FIRED COMBUSTION CHAMBER on EXISTING BOILER

SHEET 3 of 3

24

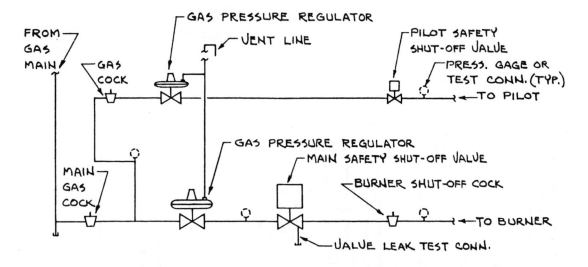

GENERAL PURPOSE

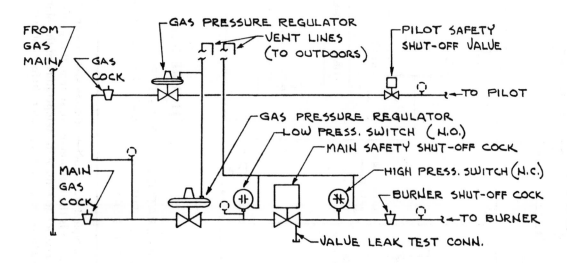

WITH GAS PRESSURE SWITCHES

**FIGURE 1-19**

TYPICAL GAS BURNER
GAS LINE TRAINS

Figure 1-19 is a typical gas burner, gas line train. It is not to be presumed to be absolutely correct in any area. The codes and various insurance regulations should be taken into account before using this detail and such changes required by codes or regulations should be incorporated in the detail.

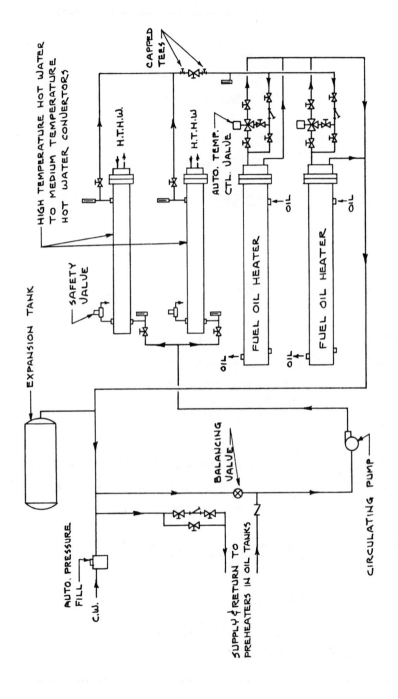

## FIGURE 1-20

HIGH TEMPERATURE HOT WATER to MEDIUM TEMPERATURE
HOT WATER FUEL OIL HEATING PUMP & HEATER SET
PIPING DIAGRAM

Figure 1-20 shows a typical No. 6 fuel oil heating system when high
temperature hot water is the medium of the system. It is generally
considered good practice to reduce the temperature of the high temperature hot water to a lower medium temperature before using it for fuel oil
heating.

26

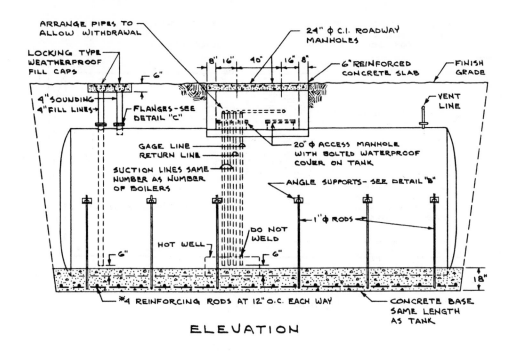

ELEVATION

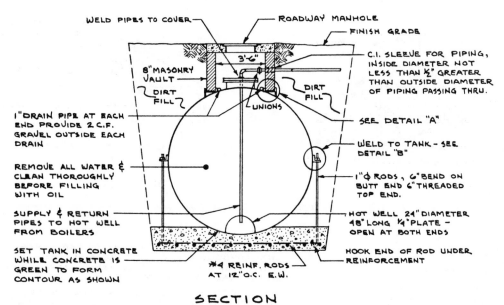

SECTION

**FIGURE 1-21**

TYPICAL UNDERGROUND INSTALLATION
OF A No. 6 OIL STORAGE TANK
(SHEET 1 OF 2)

Figures 1-21 and 1-22 are typical No. 6 fuel oil storage tank installation details, which are used except in unusual soil or underground water conditions. These details are a general standard. Actual subsurface conditions may eliminate the need for a mat.

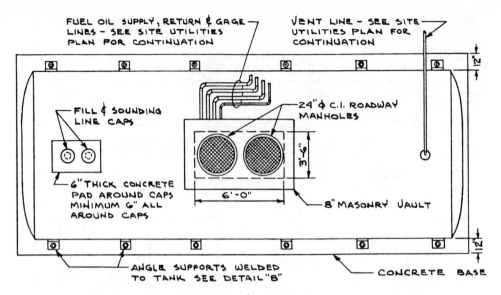

FUEL OIL SUPPLY, RETURN & GAGE LINES - SEE SITE UTILITIES PLAN FOR CONTINUATION

VENT LINE - SEE SITE UTILITIES PLAN FOR CONTINUATION

FILL & SOUNDING LINE CAPS

24"⌀ C.I. ROADWAY MANHOLES

6" THICK CONCRETE PAD AROUND CAPS MINIMUM 6" ALL AROUND CAPS

8" MASONRY VAULT

3'-6"

6'-0"

ANGLE SUPPORTS WELDED TO TANK SEE DETAIL "B"

CONCRETE BASE

PLAN VIEW

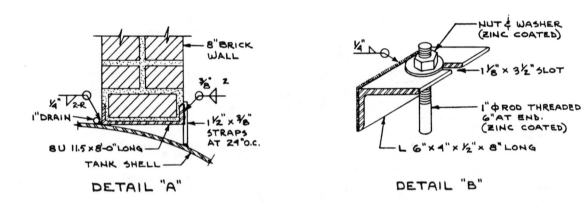

8" BRICK WALL

$\frac{3}{8}$" 2

$\frac{1}{4}$" 2-R

1" DRAIN

$1\frac{1}{2}$" x $\frac{3}{8}$" STRAPS AT 24" O.C.

8U 11.5 x 8'-0" LONG

TANK SHELL

DETAIL "A"

NUT & WASHER (ZINC COATED)

$\frac{1}{4}$"

$1\frac{1}{8}$" x $3\frac{1}{2}$" SLOT

1"⌀ ROD THREADED 6" AT END. (ZINC COATED)

L 6" x 4" x $\frac{1}{2}$" x 8" LONG

DETAIL "B"

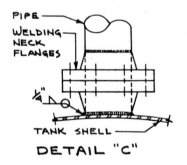

PIPE

WELDING NECK FLANGES

$\frac{1}{4}$"

TANK SHELL

DETAIL "C"

NOTES:
1. OIL STORAGE TANK SHALL BE SUITABLE FOR UNDERGROUND USE & SHALL BE BUILT IN ACCORDANCE WITH UNDERWRITERS LABORATORIES INC. & SHALL BE SO STAMPED.
2. ALL SUCTION LINES SHALL BE SUCTION TESTED TO 100 MICRONS AFTER INSTALLATION, ALL RETURN LINES SHALL BE PRESSURE TESTED TO 50 P.S.I.G. AFTER INSTALLATION.
3. ALL UNDERGROUND PIPING SHALL BE WELDED.

**FIGURE 1-22**

TYPICAL UNDERGROUND INSTALLATION OF A No. 6 OIL STORAGE TANK

(SHEET 2 OF 2)

28

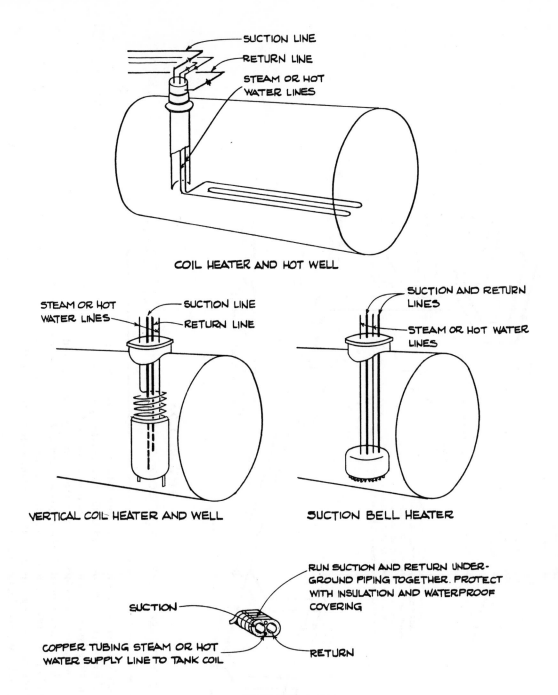

SUCTION LINE

RETURN LINE

STEAM OR HOT
WATER LINES

COIL HEATER AND HOT WELL

STEAM OR HOT
WATER LINES

SUCTION LINE

RETURN LINE

VERTICAL COIL HEATER AND WELL

SUCTION AND RETURN
LINES

STEAM OR HOT WATER
LINES

SUCTION BELL HEATER

RUN SUCTION AND RETURN UNDER-
GROUND PIPING TOGETHER. PROTECT
WITH INSULATION AND WATERPROOF
COVERING

SUCTION

COPPER TUBING STEAM OR HOT
WATER SUPPLY LINE TO TANK COIL

RETURN

## FIGURE 1-23

## TANK HEATERS

Figure 1-23 shows three different versions of heating oil in a tank. The two
common versions are the vertical coil heater and the suction bell heater.
In each case careful sizing to design conditions is required. As a general
rule, approximately 10 ft of pipe per thousand gallons of oil using 1″ steam
pipe or 1¼″ hot water pipe with the coil heater and hot well arrangement
in the top detail will provide more than enough heat. Steam should be 10
psig or greater.

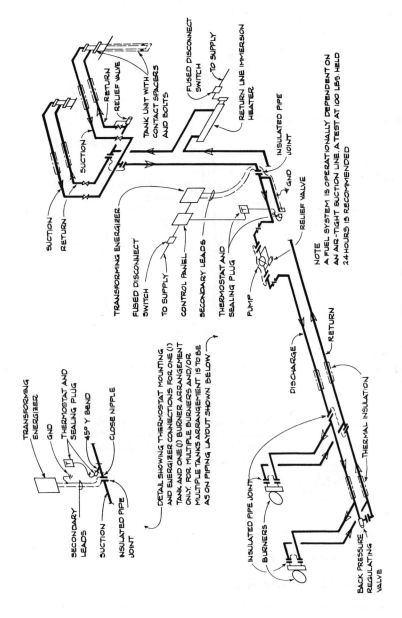

**FIGURE 1-24**

SCHEMATIC ARRANGEMENT ELECTRIC OIL PIPE HEATING SYSTEM (FOR OIL BURNERS WITH REMOTE PUMPS)

Figure 1-24 is a typical electric pipe heating detail for burners with remote pumps which is very commonly used in the New England area. The electric heating system is shown in detail. It is a type of detail that is a composite of various manufacturers recommendations and can be used pretty much as is where needed. The important part of Figure 1-24 is the location of the insulated pipe joints and the location of the secondary leads of the transforming energizer. It is also important to note that there is a relief valve bypass around the pump.

30

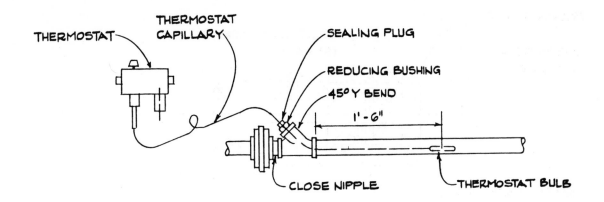

THERMOSTAT

THERMOSTAT CAPILLARY

SEALING PLUG

REDUCING BUSHING

45° Y BEND

1'-6"

CLOSE NIPPLE

THERMOSTAT BULB

## BULB MOUNTING IN PIPE RUN
## FOR ONE (1) TANK & ONE (1) BURNER ARR'G'T.

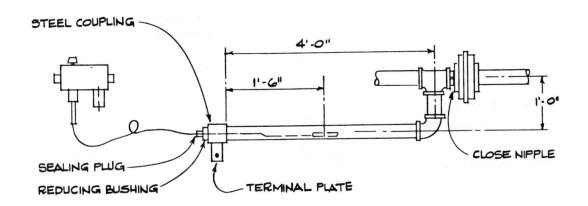

STEEL COUPLING

4'-0"

1'-6"

1'-0"

CLOSE NIPPLE

SEALING PLUG

REDUCING BUSHING

TERMINAL PLATE

## BULB MOUNTING IN PILOT STUB TAKE-OFF
## FOR MULTIPLE BURNER AND/OR MULTIPLE TANK ARR'G'T

**FIGURE 1-25**

Going along with Figure 1-24 are Figures 1-25 and 1-26. These drawings show in enlarged detail certain of the items in connection with the electric heating arrangement which have proved troublesome in the past if they were not installed properly.

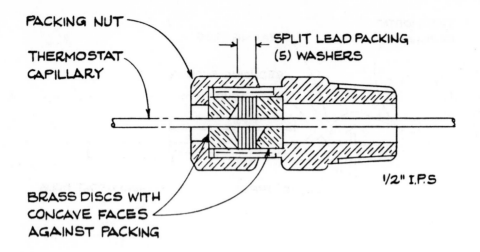

PACKING NUT

THERMOSTAT
CAPILLARY

SPLIT LEAD PACKING
(5) WASHERS

1/2" I.P.S

BRASS DISCS WITH
CONCAVE FACES
AGAINST PACKING

## SEALING PLUG DETAIL

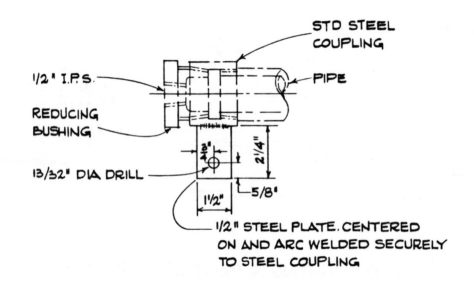

STD STEEL
COUPLING

1/2" I.P.S.

PIPE

REDUCING
BUSHING

13/32" DIA DRILL

1/2"

2 1/4"

5/8"

1 1/2"

1/2" STEEL PLATE. CENTERED
ON AND ARC WELDED SECURELY
TO STEEL COUPLING

**FIGURE 1-26**

## TERMINAL PLATE DETAIL

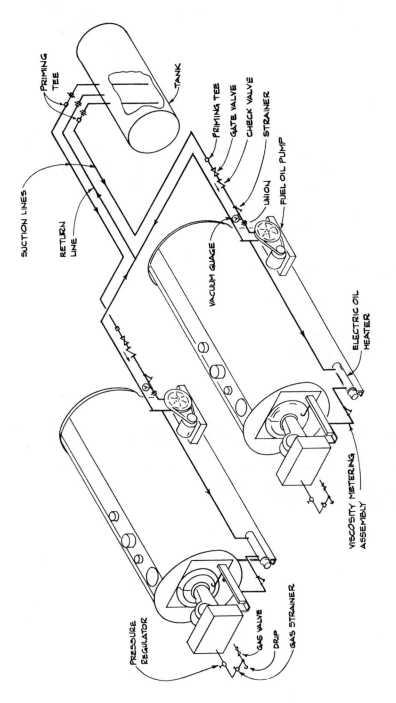

**FIGURE 1-27**

*#4 & #5 COLD FUEL OIL SYSTEM
SINGLE OR DUPLEX*

Figure 1-27 illustrates the general arrangement of a No. 4 or No. 5 fuel supply system from a single tank to one or more boilers. The detail does not show every pipe and fitting that may be necessary to complete the job for a particular project. The important point in this detail is to note the location of priming tees for each boiler and also priming tees at the tank.

33

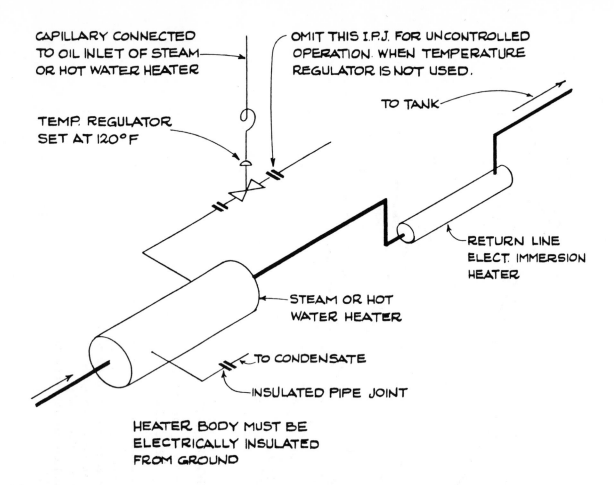

CAPILLARY CONNECTED
TO OIL INLET OF STEAM
OR HOT WATER HEATER

OMIT THIS I.P.J. FOR UNCONTROLLED
OPERATION. WHEN TEMPERATURE
REGULATOR IS NOT USED.

TO TANK

TEMP. REGULATOR
SET AT 120°F

RETURN LINE
ELECT. IMMERSION
HEATER

STEAM OR HOT
WATER HEATER

TO CONDENSATE

INSULATED PIPE JOINT

HEATER BODY MUST BE
ELECTRICALLY INSULATED
FROM GROUND

## FIGURE 1-28

# STEAM OR HOT WATER HEATERS IN RETURN LINE DETAIL

Figure 1-28 illustrates electric heaters in a steam or hot water system.
There is a special detail showing the installation of the temperature
regulator.

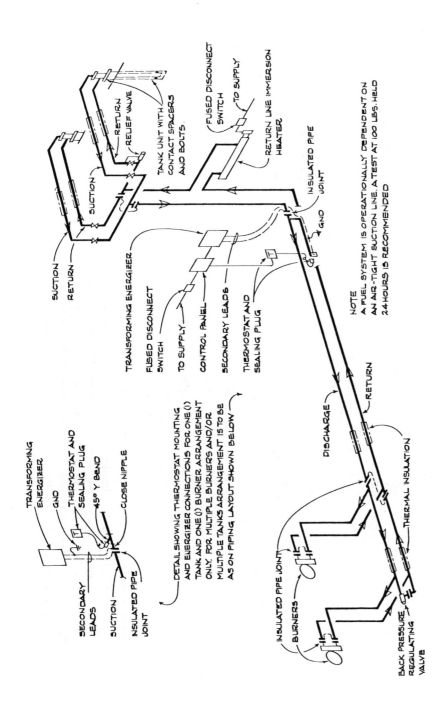

**FIGURE 1-29**

SCHEMATIC ARRANGEMENT ELECTRIC OIL PIPE HEATING SYSTEM (FOR OIL BURNERS WITH INTEGRAL PUMPS)

Figure 1-29 is in many ways similar to Figure 1-24 and it is put in this particular set of details because there are readily noted differences between the arrangement of electric oil piping heating systems when the pump is integral with the burner and no additional pump is required.

35

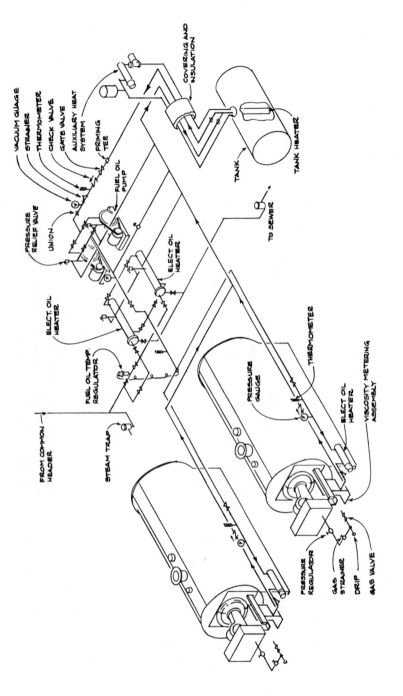

**FIGURE 1-30**

FUEL OIL SYSTEM FOR MULTIPLE UNITS
WITH COMMON PUMP #6 OIL & HOT #5 OIL

Figure 1-30 is a composite drawing which relates the installation of a system with electric heaters and remote fuel pumps. It is generally the arrangement used for No. 6 oil with a common pump or hot No. 5 oil. Again this is a detail that requires a certain amount of discretion on the part of the designer to decide exactly how this will be applied to a particular project.

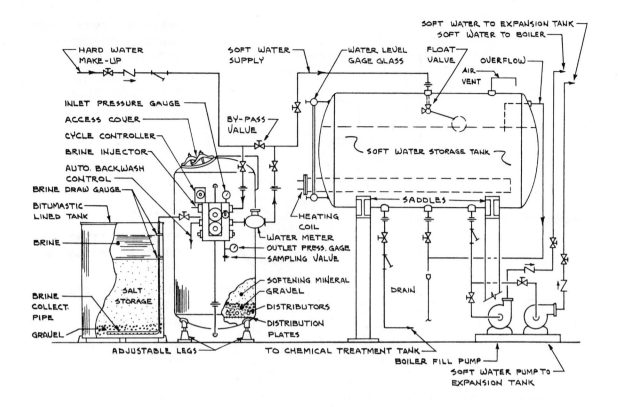

**FIGURE 1-31**

WATER SOFTENING SYSTEM FOR
HIGH TEMPERATURE HOT WATER SYSTEM

Figure 1-31 illustrates a water-softening system for a high temperature water system which, with some adjustments, could be used on a high pressure steam installation. It is again a generalized version which utilizes normal hook-ups for a particular manufacturer. Obviously, with the selection of other manufacturers there would be slight variations in the piping arrangement shown.

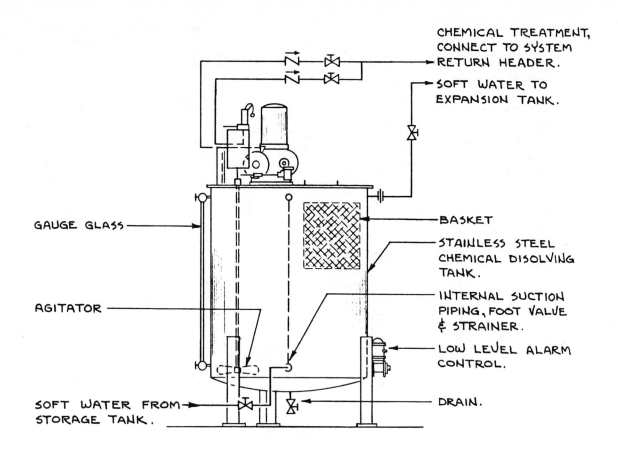

CHEMICAL TREATMENT, CONNECT TO SYSTEM RETURN HEADER.

SOFT WATER TO EXPANSION TANK.

GAUGE GLASS

AGITATOR

SOFT WATER FROM STORAGE TANK.

BASKET

STAINLESS STEEL CHEMICAL DISOLVING TANK.

INTERNAL SUCTION PIPING, FOOT VALVE & STRAINER.

LOW LEVEL ALARM CONTROL.

DRAIN.

**FIGURE 1-32**

TYPICAL HIGH TEMPERATURE HOT WATER CHEMICAL TREATMENT SYSTEM

Figure 1-32 illustrates a chemical treatment system for any high temperature hot water or high pressure steam system, which invariably has connections peculiar to the particular type of treatment. The detail shown here is for one of the relatively well-known standard chemical treatment supply systems.

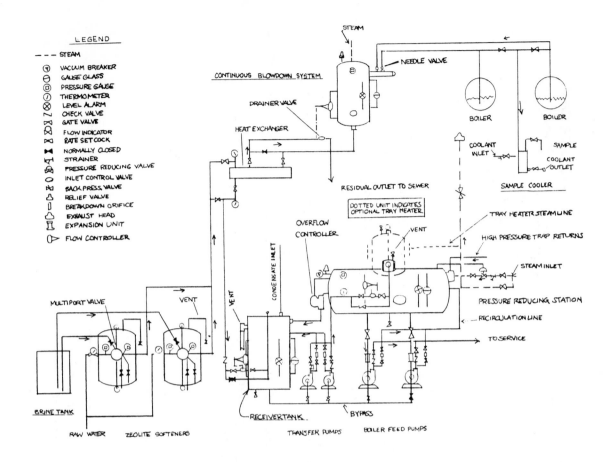

## FIGURE 1-33

BOILER FEEDWATER CONDITIONING EQUIPMENT

——— NO SCALE ———

Figure 1-33 should not be used as a detail to be directly copied. Instead it is a composite detail which can be used to check against the specific items selected and to combine them in a manner similar to the arrangement shown or present them as individual separate details. It is more of a check detail than an actual detail.

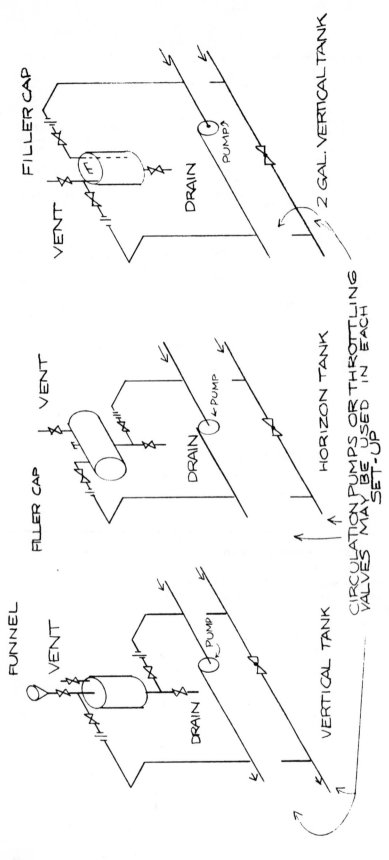

FIGURE 1-34

BY-PASS FEEDERS (TYPICAL INSTALLATION)
NO SCALE

There are various types of bypass feeders and various arrangements.
Figure 1-34 is simply an illustration of some of the relatively common
arrangements and the feeder to the system.

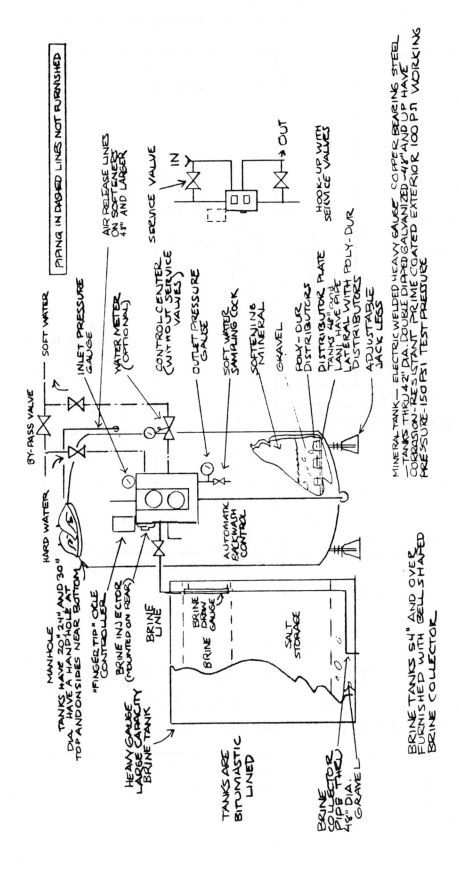

PIPING IN DASHED LINES NOT FURNISHED

AIR RELEASE LINES ON SOFTENERS 48" AND LARGER

SERVICE VALVE

IN

OUT

HOOK-UP WITH SERVICE VALVES

SOFT WATER

INLET PRESSURE GAUGE

WATER METER (OPTIONAL)

CONTROL CENTER (WITHOUT SERVICE VALVES)

OUTLET PRESSURE GAUGE

SOFT WATER SAMPLING COCK

SOFTENING MINERAL

GRAVEL

POLY-DUR DISTRIBUTORS

DISTRIBUTOR PLATE TANKS 48" AND LARGER WILL HAVE PIPE LATERAL WITH POLY-DUR DISTRIBUTORS

ADJUSTABLE JACK LEGS

BY-PASS VALVE

HARD WATER

MANHOLE
TANKS HAVE 20" 24" AND 30" DIA. HAVE A HANDHOLE AT TOP AND ON SIDES NEAR BOTTOM

"FINGER TIP" CYCLE CONTROLLER

BRINE INJECTOR (MOUNTED ON REAR)

BRINE LINE

HEAVY GAUGE LARGE CAPACITY BRINE TANK

AUTOMATIC BACKWASH CONTROL

BRINE DRAW GAUGE

BRINE

SALT STORAGE

TANKS ARE BITUMASTIC LINED

BRINE COLLECTOR PIPE THRU 48" DIA. GRAVEL

MINERAL TANK — ELECTRIC WELDED HEAVY GAUGE COPPER BEARING STEEL — TANKS THRU 42" DIA. DOUBLE DIPPED GALVANIZED — 48" AND UP HAVE CORROSION-RESISTANT PRIME COATED EXTERIOR 100 PSI WORKING PRESSURE-150 PSI TEST PRESSURE

BRINE TANKS 54" AND OVER FURNISHED WITH BELL SHAPED BRINE COLLECTOR

## FIGURE 1-35

WATER SOFTENING AND FILTERING SYSTEM

NO SCALE

Figure 1-35 again shows a water-softening and filtration system which is relatively similar to Figure 1-31 and is merely a check detail which is commonly used.

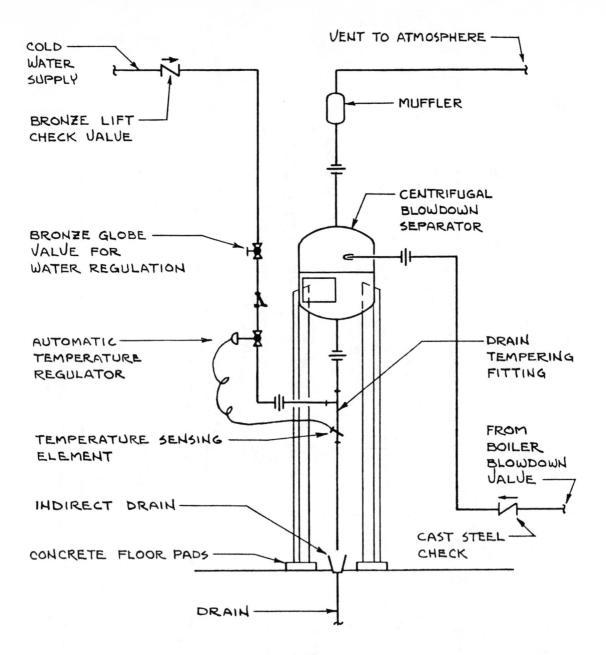

COLD WATER SUPPLY

BRONZE LIFT CHECK VALVE

BRONZE GLOBE VALVE FOR WATER REGULATION

AUTOMATIC TEMPERATURE REGULATOR

TEMPERATURE SENSING ELEMENT

INDIRECT DRAIN

CONCRETE FLOOR PADS

DRAIN

VENT TO ATMOSPHERE

MUFFLER

CENTRIFUGAL BLOWDOWN SEPARATOR

DRAIN TEMPERING FITTING

FROM BOILER BLOWDOWN VALVE

CAST STEEL CHECK

**FIGURE 1-36**

TYPICAL HIGH TEMPERATURE HOT WATER BOILER BLOWDOWN PIPING FOR CENTRIFUGAL SEPARATOR

Figure 1-36 is a standard detail of a high temperature hot water boiler blowdown piping installation. There is very little to make this detail stand out. Perhaps the important thing is to carefully size the indirect drain in order to provide sufficient capacity for the blowdown steam that occurs.

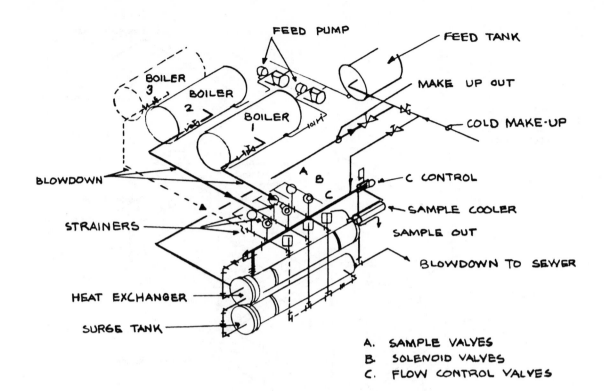

FEED PUMP

FEED TANK

BOILER 3

BOILER 2

BOILER 1

MAKE UP OUT

COLD MAKE-UP

BLOWDOWN

A B C

C CONTROL

SAMPLE COOLER

STRAINERS

SAMPLE OUT

BLOWDOWN TO SEWER

HEAT EXCHANGER

SURGE TANK

A. SAMPLE VALVES
B. SOLENOID VALVES
C. FLOW CONTROL VALVES

## FIGURE 1-37

## AUTOMATIC APPORTIONING OF BLOW DOWN TO MAKE UP WATER

Figure 1-37 is not a detail to be used directly, but rather a detail to be used as a check on the arrangement needed to make the actual detail for apportioning of blowdown to make-up water.

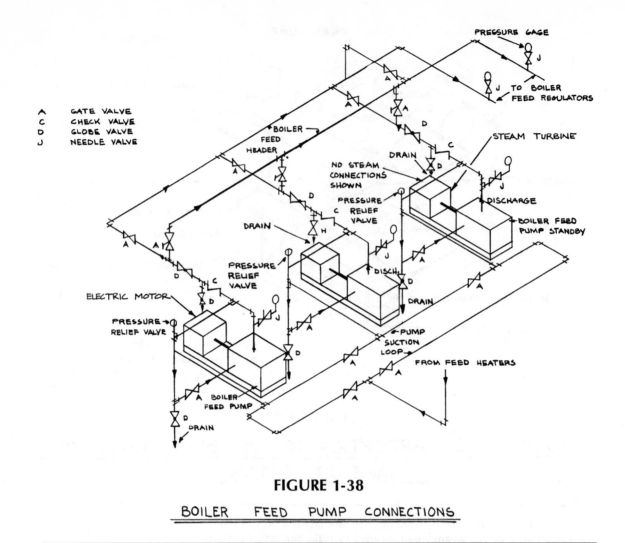

A GATE VALVE
C CHECK VALVE
D GLOBE VALVE
J NEEDLE VALVE

PRESSURE GAGE

J

J TO BOILER
FEED REGULATORS

STEAM TURBINE

BOILER
FEED
HEADER

NO STEAM
CONNECTIONS
SHOWN

DRAIN

C

DISCHARGE

BOILER FEED
PUMP STANDBY

PRESSURE
RELIEF
VALVE

DRAIN

PRESSURE
RELIEF
VALVE

DISCH.

DRAIN

ELECTRIC MOTOR

PUMP
SUCTION
LOOP

PRESSURE
RELIEF VALVE

FROM FEED HEATERS

BOILER
FEED PUMP

DRAIN

## FIGURE 1-38

BOILER   FEED   PUMP   CONNECTIONS

Figure 1-38 is a relatively standard detail showing two electric pumps and one steam driven pump supplying boiler feed water for a 125-lb system. The important thing to note is the bypass line and valve situation around each of the pumps. Normally in this line either some sort of pressure control or usually an orifice is inserted, which allows a certain amount of water to continuously bypass.

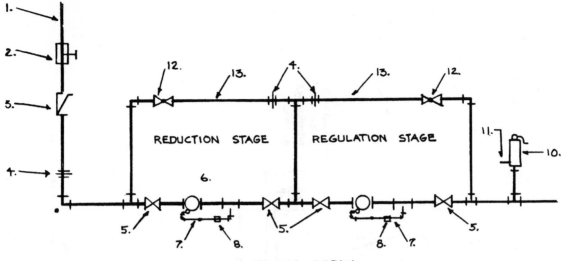

PLAN VIEW

1. SERVICE PIPE
2. SERVICE VALVE
3. STRAINER
4. FLANGED UNION
5. GATE VALVE
6. PILOT-OPERATED PRESSURE-REDUCING VALVE
7. BALANCE LINE
8. PRESSURE GAGE
9. PILOT-OPERATED PRESSURE-REGULATING VALVE
10. PRESSURE RELIEF VALVE (OPTIONAL)
11. VENT
12. GLOBE VALVE
13. BY-PASS LINE

## FIGURE 1-39

TWO STAGE PRESSURE-REDUCING & PRESSURE-REGULATING STATION EQUIPMENT. USED WHERE HIGH-PRESSURE STEAM IS SUPPLIED FOR LOW-PRESSURE REQUIREMENTS.

Figure 1-39 is familiar to every engineer and it is literally the perfect example to show why this book was written. The detail is common, simple, and easy to produce. It is a detail that can easily be copied.

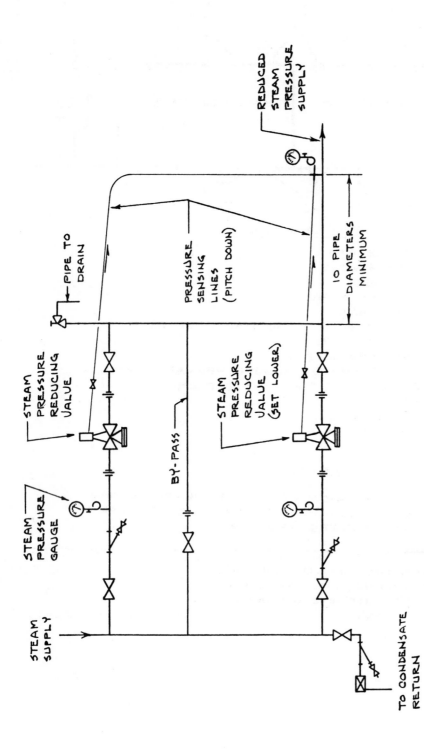

**FIGURE 1-40**

PARALLEL INSTALLING
of STEAM PRESSURE REDUCING VALVES

Figure 1-40 is similar in premise to Figure 1-39. It is a standard arrangement for the parallel installation of pressure-reducing valves.

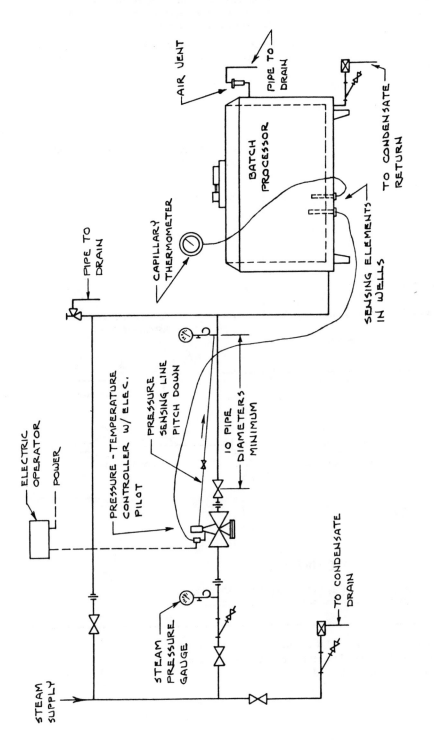

**FIGURE 1-41**

PRESSURE REDUCING VALVE
FOR BATCH PROCESSING APPLICATION

Figure 1-41 is not of itself a very intricate detail. The usual run of batch processors will require the sensing element to be in the bottom rather than the side, top, or some other location. It is for this reason that the detail was created.

47

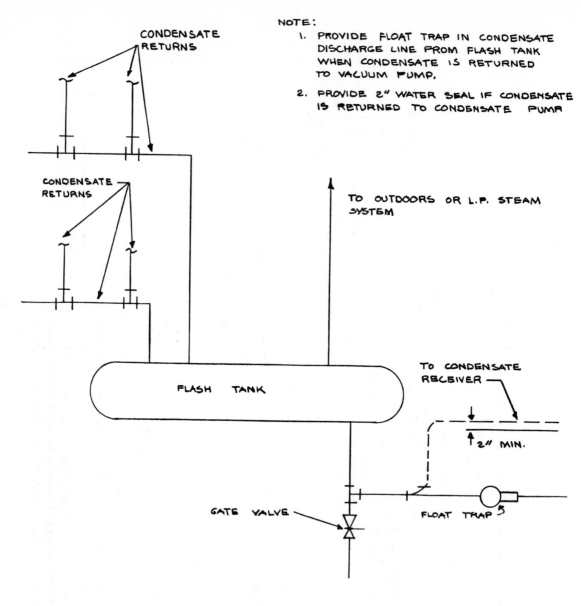

**FIGURE 1-42**

TYPICAL FLASH TANK PIPING

Figure 1-42 shows typical flash tank piping and it is one version of this type of piping which is relatively standard. The notes on the detail are of considerable importance and should be followed.

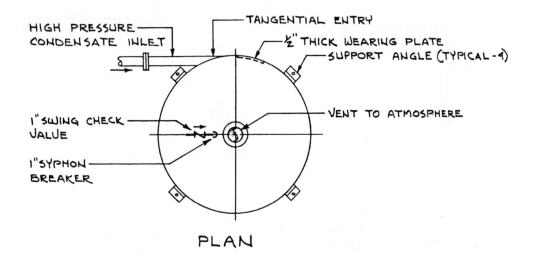

HIGH PRESSURE CONDENSATE INLET

TANGENTIAL ENTRY

½" THICK WEARING PLATE

SUPPORT ANGLE (TYPICAL - 4)

1" SWING CHECK VALVE

VENT TO ATMOSPHERE

1" SYPHON BREAKER

PLAN

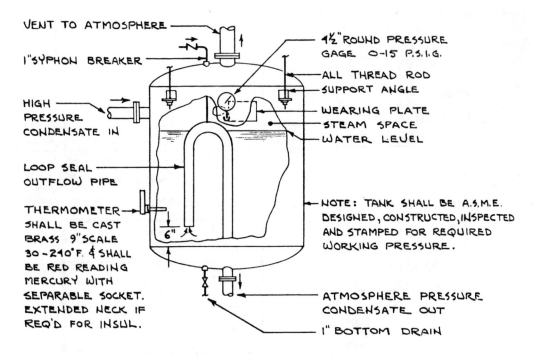

VENT TO ATMOSPHERE

1" SYPHON BREAKER

4½" ROUND PRESSURE GAGE 0-15 P.S.I.G.

ALL THREAD ROD

SUPPORT ANGLE

HIGH PRESSURE CONDENSATE IN

WEARING PLATE

STEAM SPACE

WATER LEVEL

LOOP SEAL OUTFLOW PIPE

THERMOMETER SHALL BE CAST BRASS 9" SCALE 30 - 210° F. & SHALL BE RED READING MERCURY WITH SEPARABLE SOCKET. EXTENDED NECK IF REQ'D FOR INSUL.

6"

NOTE: TANK SHALL BE A.S.M.E. DESIGNED, CONSTRUCTED, INSPECTED AND STAMPED FOR REQUIRED WORKING PRESSURE.

ATMOSPHERE PRESSURE CONDENSATE OUT

1" BOTTOM DRAIN

ELEVATION

**FIGURE 1-43**

INSTALLATION OF FLASH TANK

Figure 1-43 is a relatively standard version of the vertical flash tank and, while the detail itself is not unusual in any way, it is very important to include the loop seal shown in the detail to ensure that atmospheric pressure condensate is the liquid being delivered to the system drain.

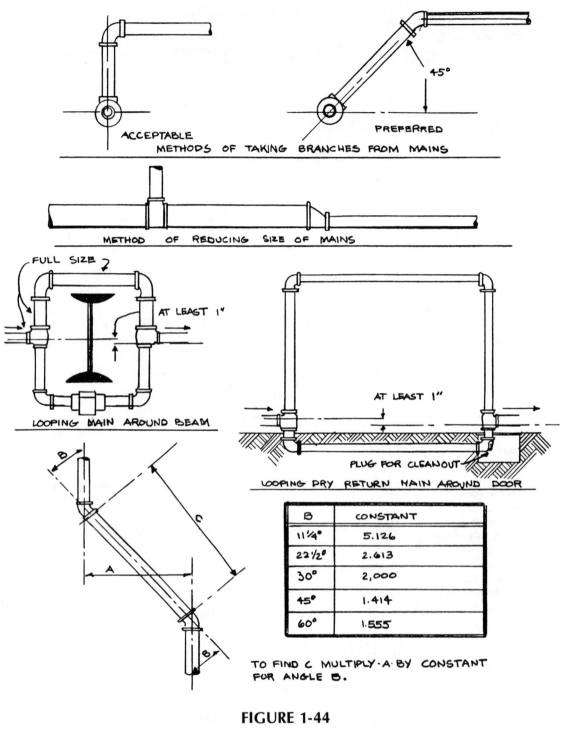

ACCEPTABLE

PREFERRED

45°

METHODS OF TAKING BRANCHES FROM MAINS

METHOD OF REDUCING SIZE OF MAINS

FULL SIZE

AT LEAST 1"

LOOPING MAIN AROUND BEAM

AT LEAST 1"

PLUG FOR CLEANOUT

LOOPING DRY RETURN MAIN AROUND DOOR

| B | CONSTANT |
|-------|----------|
| 11¼° | 5.126 |
| 22½° | 2.613 |
| 30° | 2,000 |
| 45° | 1.414 |
| 60° | 1.555 |

TO FIND C MULTIPLY A BY CONSTANT FOR ANGLE B.

**FIGURE 1-44**

SPECIAL PIPE DETAILS

Figures 1-44 and 1-45 are seemingly rather simple details of piping and risers. In practice, these little points are of such importance and are the source of such frustration that it is essential that the details be used whenever they apply.

50

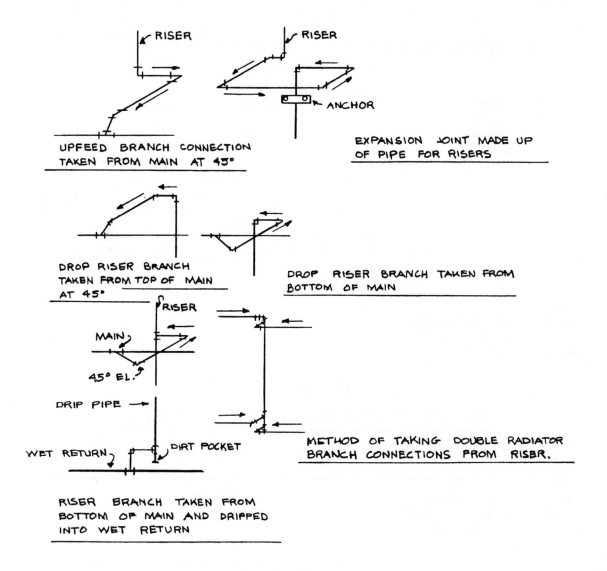

RISER

RISER

ANCHOR

UPFEED BRANCH CONNECTION
TAKEN FROM MAIN AT 45°

EXPANSION JOINT MADE UP
OF PIPE FOR RISERS

DROP RISER BRANCH
TAKEN FROM TOP OF MAIN
AT 45°

DROP RISER BRANCH TAKEN FROM
BOTTOM OF MAIN

RISER

MAIN

45° EL.

DRIP PIPE

WET RETURN

DIRT POCKET

METHOD OF TAKING DOUBLE RADIATOR
BRANCH CONNECTIONS FROM RISER.

RISER BRANCH TAKEN FROM
BOTTOM OF MAIN AND DRIPPED
INTO WET RETURN

**FIGURE 1-45**

RISER DETAILS

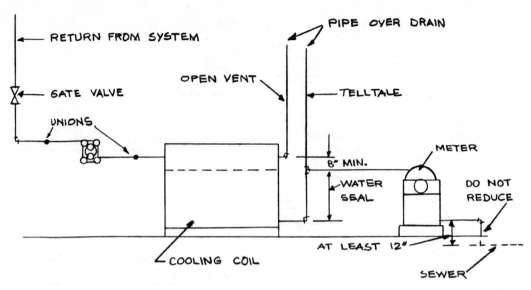

RETURN FROM SYSTEM

PIPE OVER DRAIN

OPEN VENT

GATE VALVE

TELLTALE

UNIONS

METER

8" MIN.

WATER SEAL

DO NOT REDUCE

AT LEAST 12"

SEWER

COOLING COIL

NOTE:
RANGE BOILER, EXPANSION
TANK OR RADIATOR MAY BE
USED FOR COOLING COIL.

## FIGURE 1-46

## METER INSTALLED GRAVITY SYSTEM

### FOR PURCHASED STEAM

Figure 1-46 illustrates a standard way of measuring steam consumption by use of a gravity meter. The detail is relatively simple and it is quite good in its particular type of application.

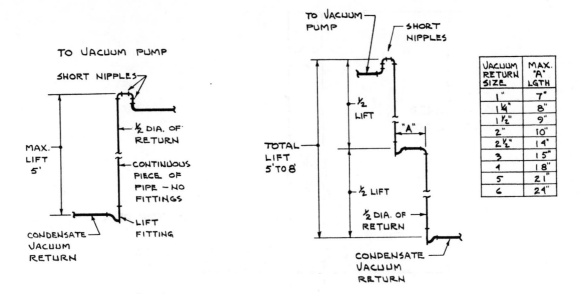

ONE-STEP CONDENSATE LIFT          TWO-STEP CONDENSATE LIFT

| VACUUM RETURN SIZE | MAX. "A" LGTH |
|---|---|
| 1" | 7" |
| 1¼" | 8" |
| 1½" | 9" |
| 2" | 10" |
| 2½" | 14" |
| 3 | 15" |
| 4 | 18" |
| 5 | 21" |
| 6 | 24" |

**FIGURE 1-47**

Figures 1-47 and 1-48 are piping details that illustrate certain very troublesome points in any piping system. It is essential that these details be followed whenever they apply.

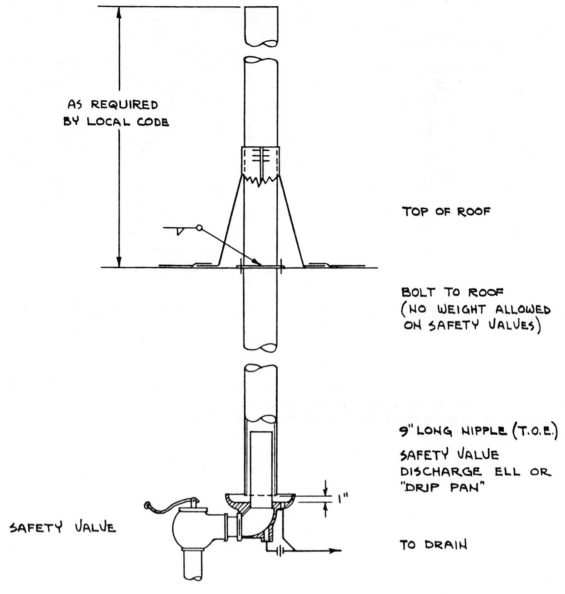

AS REQUIRED
BY LOCAL CODE

TOP OF ROOF

BOLT TO ROOF
(NO WEIGHT ALLOWED
ON SAFETY VALVES)

9" LONG NIPPLE (T.O.E.)

SAFETY VALVE
DISCHARGE ELL OR
"DRIP PAN"

1"

SAFETY VALVE

TO DRAIN

NIPPLE AS SHORT AS POSSIBLE

### FIGURE 1-48

## TYPICAL HIGH PRESSURE BOILER
## SAFETY VALVE DISCHARGE PIPING

(HIGH PRESSURE BOILERS I.E. STEAM BOILERS
OVER 15 P.S.I. AND HOT WATER BOILERS OVER
160 P.S.I. OR OPERATING OVER 250°F.)

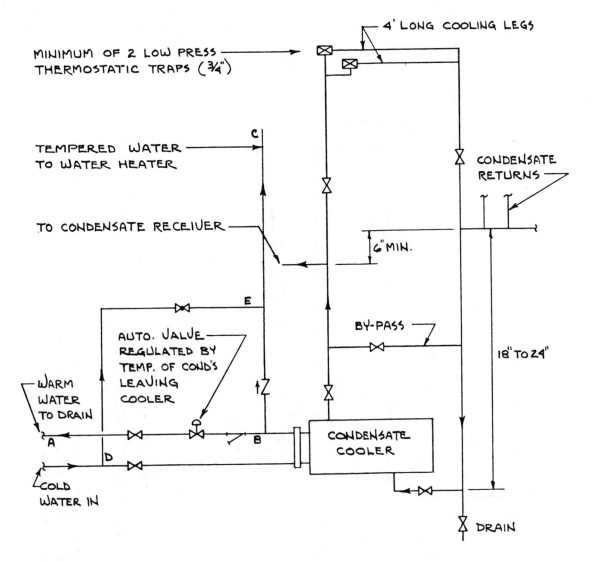

MINIMUM OF 2 LOW PRESS — THERMOSTATIC TRAPS (¾")

4' LONG COOLING LEGS

TEMPERED WATER TO WATER HEATER

CONDENSATE RETURNS

TO CONDENSATE RECEIVER

6" MIN.

AUTO. VALVE REGULATED BY TEMP. OF COND'S LEAVING COOLER

BY-PASS

18" TO 24"

WARM WATER TO DRAIN

CONDENSATE COOLER

COLD WATER IN

DRAIN

NOTES:

1. DELETE SECTION A-B IF CONTROL OF LEAVING CONDENSATE TEMPERATURE IS NOT REQUIRED

2. DELETE SECTIONS B-C & D-E IF PRE HEATING OF HOT WATER IS NOT REQUIRED

**FIGURE 1-49**

TYPICAL STEAM CONDENSATE COOLER PIPING

Figure 1-49 is a typical steam condensate cooler piping. It is a system for further cooling of condensate prior to returning to the system for whatever reason condensate cooling may be required. The detail is fairly complete and unless there are special reasons it is to be used as is.

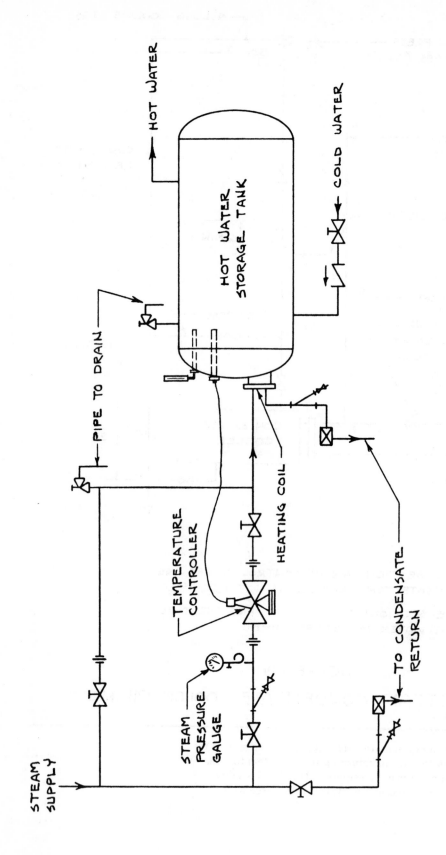

**FIGURE 1-50**

STEAM HEAT TEMPERATURE
CONTROL OF HOT WATER HEATER

Figure 1-50 is an old standby drawing in which not every fitting is shown.
It can be used fairly well on an as is basis with sizes and the missing
fittings shown as required.

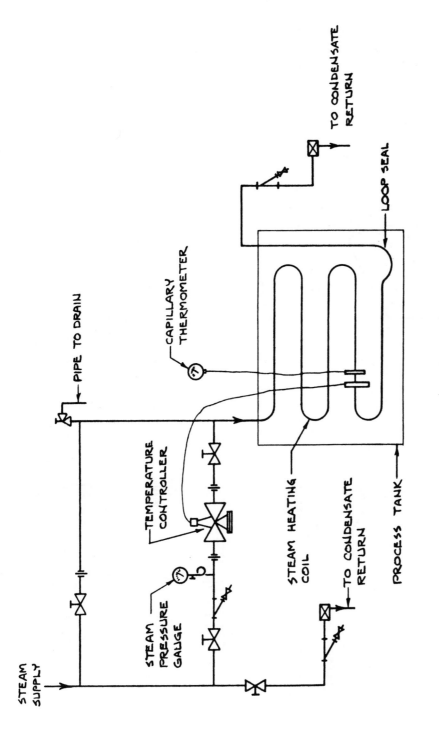

**FIGURE 1-51**

STEAM HEAT TEMPERATURE
CONTROL of PROCESS TANK

Figure 1-51 is an illustration of control of steam to a process tank. The important thing is the location of the sensing device for the temperature controller and the thermometer.

57

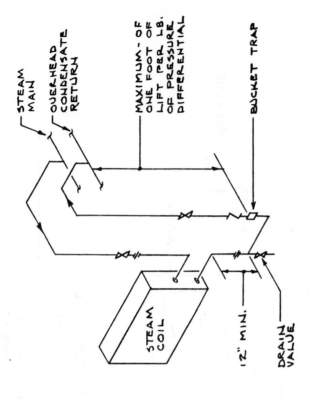

STEAM MAIN

OVERHEAD CONDENSATE RETURN

MAXIMUM - OF ONE FOOT OF LIFT PER LB. OF PRESSURE DIFFERENTIAL

BUCKET TRAP

STEAM COIL

12" MIN.

DRAIN VALUE

MULTIPLE COIL
HIGH PRESSURE PIPING

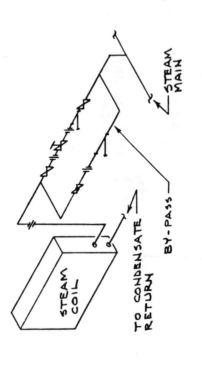

( BY-PASS SAME SIZE AS CONTROL VALVE PORT BUT NOT LESS THAN ½".)

STEAM MAIN

STEAM COIL

BY-PASS

TO CONDENSATE RETURN

HIGH or MEDIUM PRESSURE
COIL PIPING

**FIGURE 1-52**

Figure 1-52 illustrates high or medium pressure individual coil piping as well as high pressure multiple coil piping. The important thing to note in this drawing is the difference in elevation between the connection at the coil and the horizontal run to the condensate return line. Twelve inches is the minimum for the drop in this particular situation in order to insure the proper flow of condensate to the trap and the condensate return system.

58

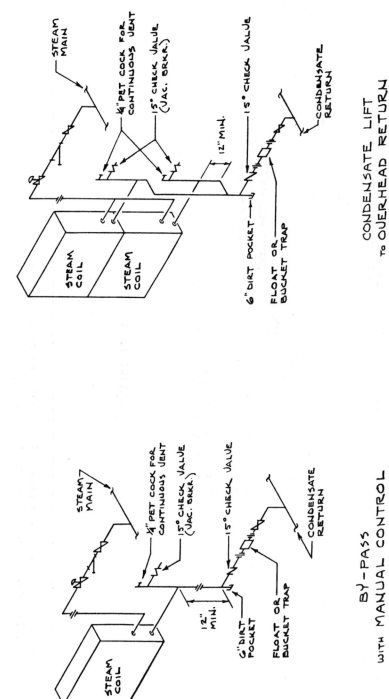

CONDENSATE LIFT
to OVERHEAD RETURN

BY-PASS
WITH MANUAL CONTROL

**FIGURE 1-53**

Figure 1-53 is a set of details illustrating two points that frequently prove troublesome. Designers should carefully note on the plans the takeoff of the mains, the 12" minimum into the trap, and the maximum of 1 ft of lift per pound of differential shown between the overhead condensate and the condensate out of the trap.

59

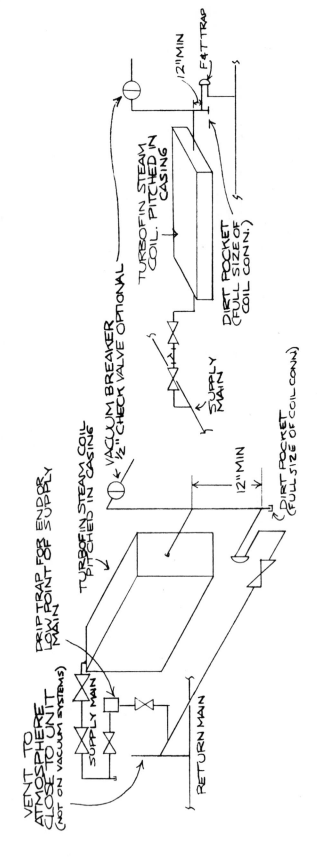

**LEGEND**

⋈ GATE VALVE
⋈ CONTROL VALVE
⊸ FLOAT & THERMOSTATIC TRAP
⊸ CHECK VALVE
⊶ STRAINER

VENT TO ATMOSPHERE CLOSE TO UNIT (NOT ON VACUUM SYSTEMS)

DRIP TRAP FOR END OR LOW POINT OF SUPPLY MAIN

TURBOFIN STEAM COIL PITCHED IN CASING

SUPPLY MAIN

VACUUM BREAKER ½" CHECK VALVE OPTIONAL

RETURN MAIN

12" MIN

(DIRT POCKET (FULL SIZE OF COIL CONN.)

TURBOFIN STEAM COIL PITCHED IN CASING

SUPPLY MAIN

DIRT POCKET (FULL SIZE OF COIL CONN.)

12" MIN

F&T TRAP

**FIGURE 1-54**

PIPING ARRANGEMENTS
NO SCALE

Figure 1-54 is a sketch arrangement design of the system to be used to visualize all of the connections and their locations.

60

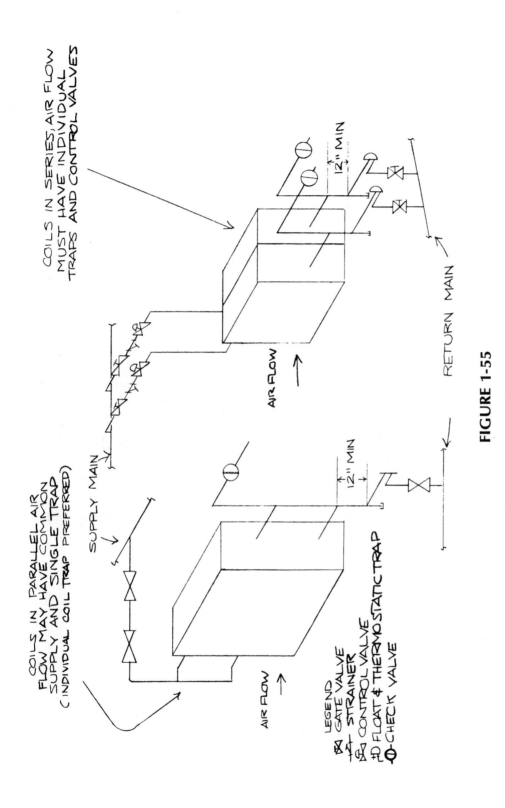

COILS IN SERIES, AIR FLOW
MUST HAVE INDIVIDUAL
TRAPS AND CONTROL VALVES

12" MIN

COILS IN PARALLEL AIR
FLOW MAY HAVE COMMON
SUPPLY AND SINGLE TRAP
( INDIVIDUAL COIL TRAP PREFERRED)

SUPPLY MAIN

AIR FLOW

RETURN MAIN

AIR FLOW

12" MIN

LEGEND
GATE VALVE
STRAINER
CONTROL VALVE
FLOAT & THERMOSTATIC TRAP
CHECK VALVE

**FIGURE 1-55**

SINGLE COIL VERTICAL AIR FLOW
NO SCALE

Figure 1-55 is another version of this same idea of preliminary sketching
of piping connections.

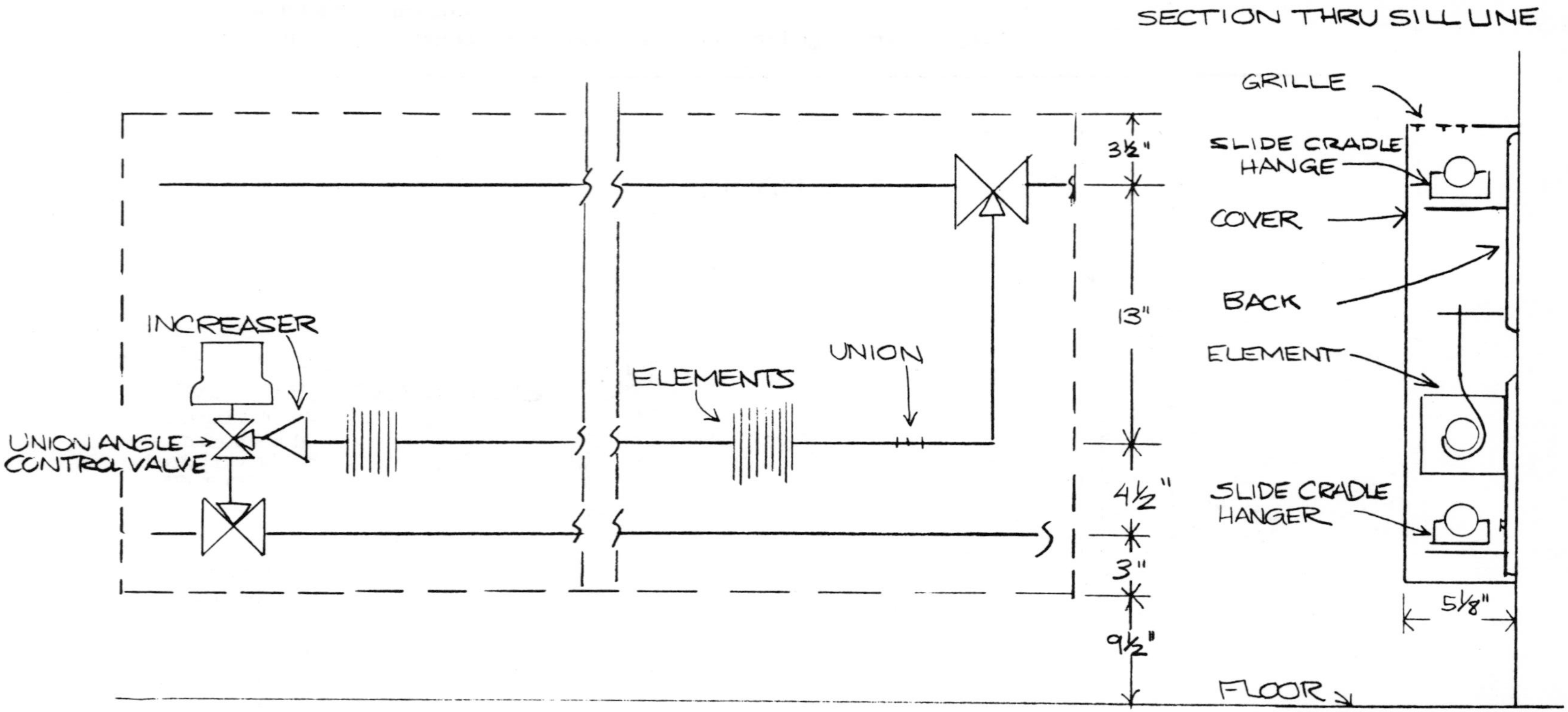

**FIGURE 1-56**

SILL LINE PIPING DIAGRAM
WITH SUPPLY AND RETURN MAINS IN COVER
NO SCALE

Figure 1-56 is a standard arrangement of single element in radiation with supply and return hot water piping in the same cover. Note that the supply piping is below the element and the return piping is above.

62

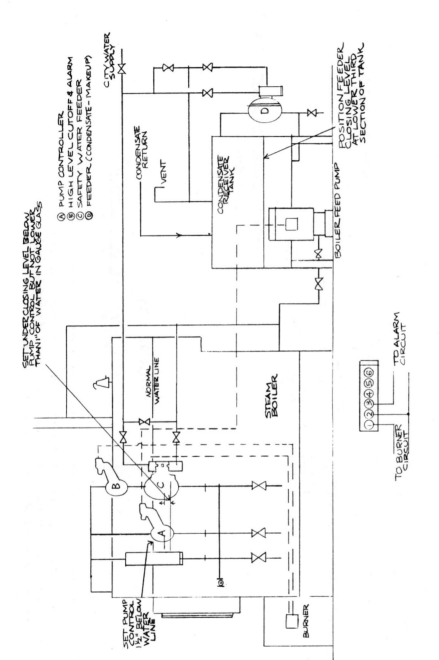

**FIGURE 1-57**

PUMP CONTROLLER, HIGH LEVEL CUT OFF & ALARM, SAFETY
WATER FEEDER AND CONDENSATE FEEDER FOR CLOSED HEATING SYSTEMS
— NO SCALE —

Figure 1-57 is another sketch drawing to be used to establish a relationship between controllers on a unit system and how a generalized arrangement will occur. It is also a detail to visualize the details required.

63

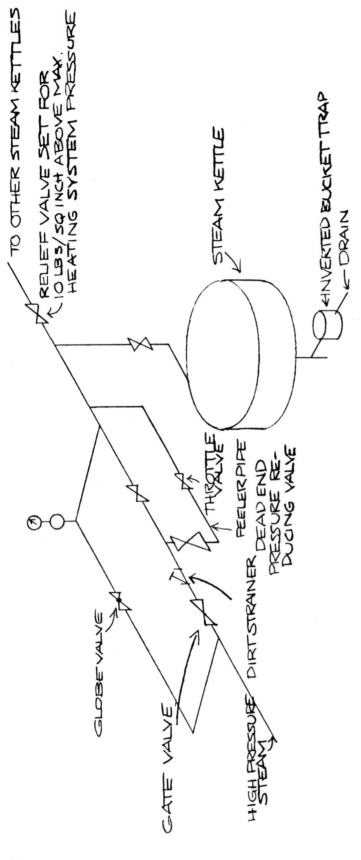

**FIGURE 1-58**

TYPICAL INSTALLATION FOR STEAM KETTLE
NO SCALE

As in other drawings, Figure 1-58 is another sketch of steam piping to a steam kettle and illustrates one of the standard ways of supplying high pressure steam and relief valves where needed in the system.

64

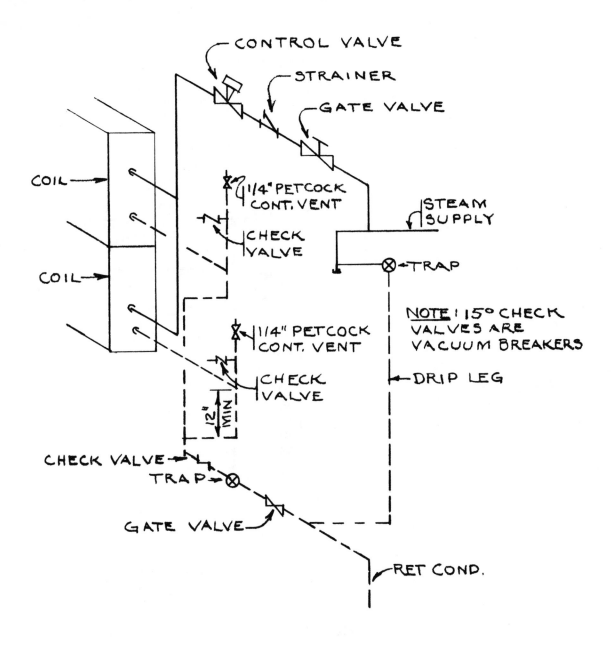

**FIGURE 1-59**

## Non-Freeze
## High Pressure System (Up to 200 Psig)
— No Scale —

Figure 1-59 is similar to previous drawings and sketches the connecting high pressure returns to steam coils. In this case it is nonfreeze. Attention is called to the ¼" pet cock put into the system for continuous vent and the check valve for vacuum-breaking purposes.

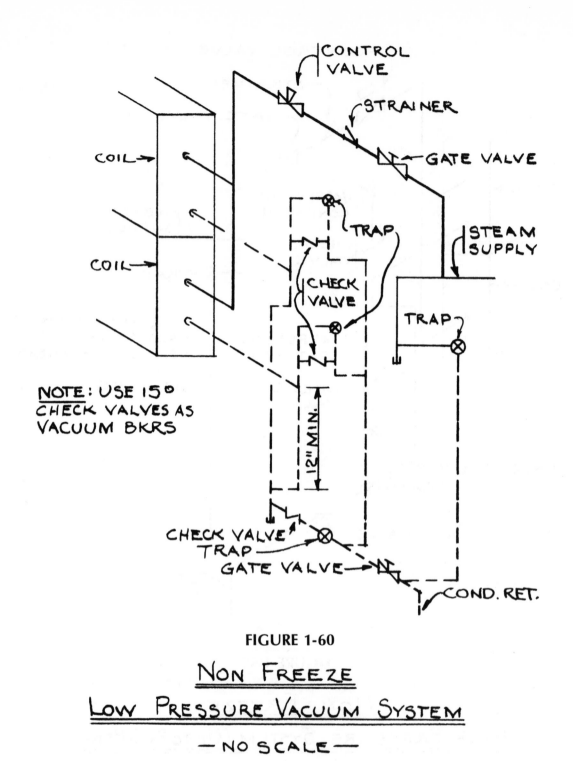

CONTROL VALVE

STRAINER

GATE VALVE

COIL

COIL

STEAM SUPPLY

TRAP

CHECK VALVE

TRAP

NOTE: USE 15° CHECK VALVES AS VACUUM BKRS

12" MIN.

CHECK VALVE

TRAP

GATE VALVE

COND. RET.

**FIGURE 1-60**

## Non Freeze

## Low Pressure Vacuum System

— NO SCALE —

Figure 1-60 is similar to Figure 1-59, but for a low pressure situation.

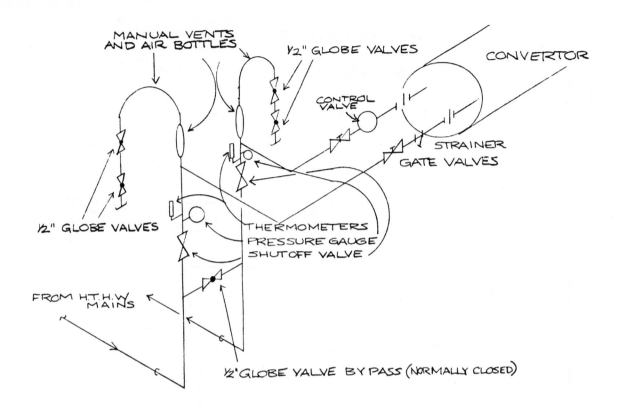

MANUAL VENTS
AND AIR BOTTLES

½" GLOBE VALVES

CONVERTOR

CONTROL
VALVE

STRAINER

GATE VALVES

½" GLOBE VALVES

THERMOMETERS
PRESSURE GAUGE
SHUTOFF VALVE

FROM H.T.H.W.
MAINS

½" GLOBE VALVE BY PASS (NORMALLY CLOSED)

INDIVIDUAL BUILDING PIPING
HTHW CONVERTOR INSTALLATION

**FIGURE 1-61**

BUILDING PIPING CONNECTIONS
——— NO SCALE ———

Figure 1-61 is certainly not unique in terms of piping arrangement. However, it is important to note that in a high temperature hot water distribution system it is necessary to have vents on the system and the arrangement shown with approximately 12″ radius turns 180° down has proven to be highly satisfactory for venting purposes.

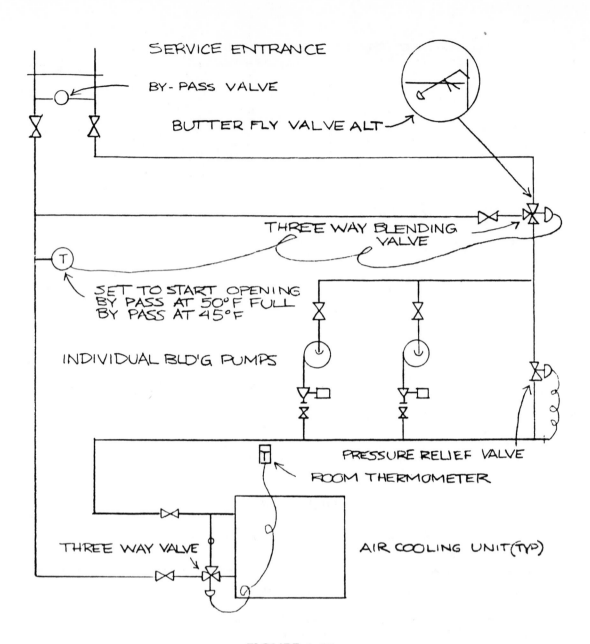

SERVICE ENTRANCE

BY-PASS VALVE

BUTTER FLY VALVE ALT

THREE WAY BLENDING VALVE

T

SET TO START OPENING BY PASS AT 50°F FULL BY PASS AT 45°F

INDIVIDUAL BLD'G PUMPS

PRESSURE RELIEF VALVE

ROOM THERMOMETER

THREE WAY VALVE

AIR COOLING UNIT(TYP)

**FIGURE 1-62**

INDIVIDUAL BUILDING PIPING
CONSTANT VOLUME — VARIABLE TEMPERATURE
CHILLED WATER

CHILLED WATER SYSTEM
—————— NO SCALE ——————

Figures 1-62 and 1-63 are not details to be copied. They are schematic arrangements of central chilled water distribution where using the building does or does not have a three way valve and the desire of the system designer is to maintain constant volume. The generalized sketches shown should be a part of the overall piping schematic of the overall distribution system and should be incorporated selectively as required by the individual designer.

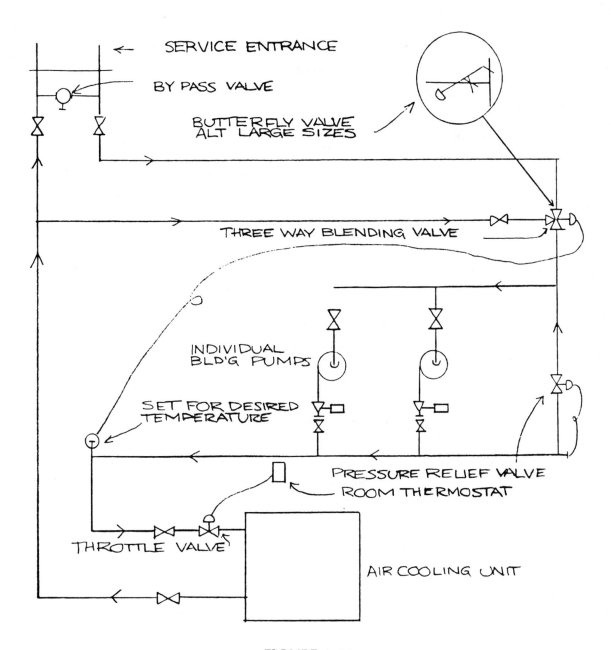

SERVICE ENTRANCE

BY PASS VALVE

BUTTERFLY VALVE
ALT LARGE SIZES

THREE WAY BLENDING VALVE

INDIVIDUAL
BLD'G PUMPS

SET FOR DESIRED
TEMPERATURE

PRESSURE RELIEF VALVE
ROOM THERMOSTAT

THROTTLE VALVE

AIR COOLING UNIT

**FIGURE 1-63**

INDIVIDUAL BUILDING PIPING
SELECTIVE TEMPERATURE — CONSTANT VOLUME

CHILLED WATER SYSTEM
NO SCALE

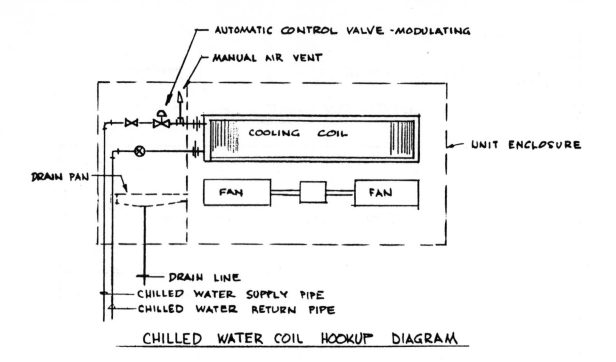

AUTOMATIC CONTROL VALVE - MODULATING

MANUAL AIR VENT

COOLING COIL

UNIT ENCLOSURE

DRAIN PAN

FAN          FAN

DRAIN LINE
CHILLED WATER SUPPLY PIPE
CHILLED WATER RETURN PIPE

CHILLED WATER COIL HOOKUP DIAGRAM

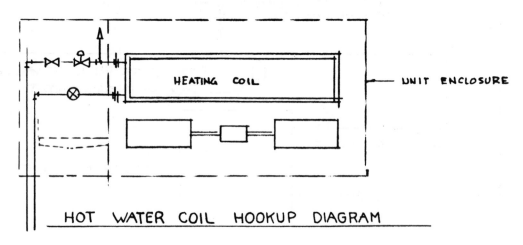

HEATING COIL

UNIT ENCLOSURE

HOT WATER COIL HOOKUP DIAGRAM

**FIGURE 1-64**

INSTALLATION OF DUAL COIL FAN COIL UNIT

NO SCALE

Figure 1-64 is a drawing that is shown in a four pipe system of a dual fan coil unit which is frequently used and which can be more easily presented as two separate details rather than try to show all of the piping in a single detail.

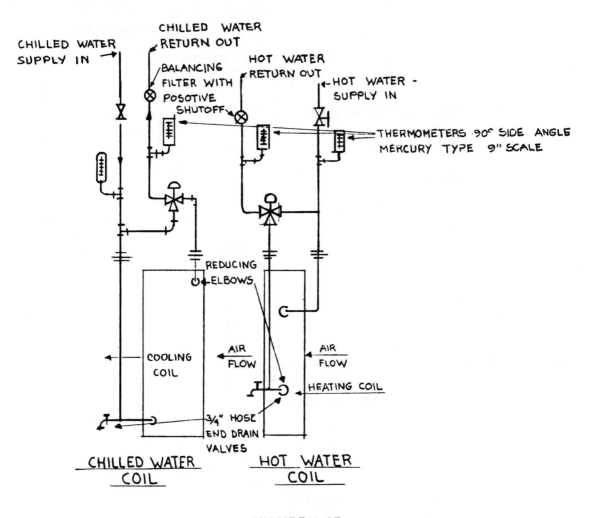

CHILLED WATER
SUPPLY IN

CHILLED WATER
RETURN OUT

BALANCING
FILTER WITH
POSOTIVE
SHUTOFF

HOT WATER
RETURN OUT

HOT WATER -
SUPPLY IN

THERMOMETERS 90° SIDE ANGLE
MERCURY TYPE 9" SCALE

REDUCING
ELBOWS

COOLING
COIL

AIR
FLOW

AIR
FLOW

HEATING COIL

¾" HOSE
END DRAIN
VALVES

CHILLED WATER
COIL

HOT WATER
COIL

**FIGURE 1-65**

COIL HOOKUP

NO SCALE

Figure 1-65 is a standard version of chilled and hot water coil piping
through an air handling unit containing both a heating and cooling coil
with balancing valves and thermometers carefully noted.

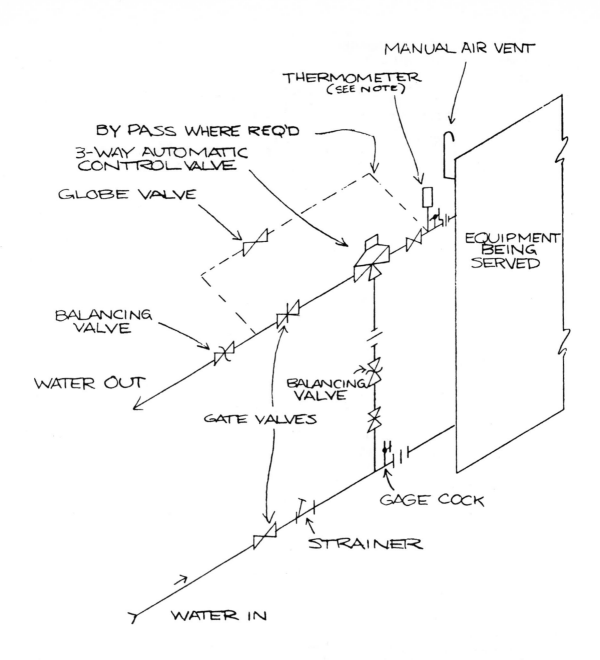

MANUAL AIR VENT

THERMOMETER
(SEE NOTE)

BY PASS WHERE REQ'D

3-WAY AUTOMATIC
CONTROL VALVE

GLOBE VALVE

EQUIPMENT
BEING
SERVED

BALANCING
VALVE

WATER OUT

BALANCING
VALVE

GATE VALVES

GAGE COCK

STRAINER

WATER IN

NOTE: INSTALLATION OF THERMOMETERS
SHOULD NOT RESTRICT WATER FLOW

**FIGURE 1-66**

TYPICAL HOT WATER AUTOMATIC CONTROL
VALVE PIPING WITH 3-WAY VALVE

Figure 1-66 is a standard sketch drawing. The important part of this
drawing is the note that the installation of the thermometer should not
restrict water flow.

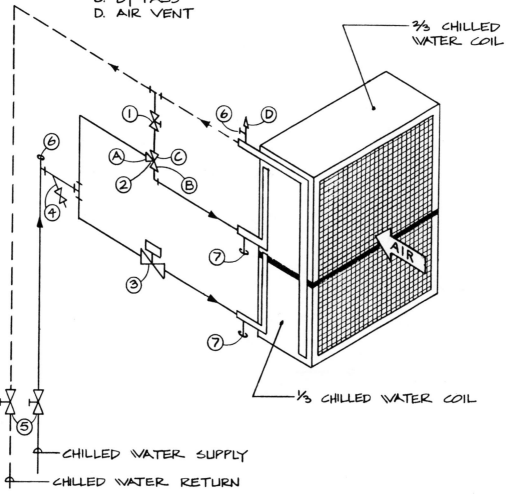

## LEGEND

1. BALANCING VALVE
2. 3-WAY MODULATING VALVE
3. SOLENOID VALVE
4. STRAINER WITH BLOW-OFF VALVE
5. BALANCING AND STOP
6. SHRADER FITTINGS FOR PRESSURE GAUGES
7. ¼" DRAIN
A. INLET
B. COIL
C. BY-PASS
D. AIR VENT

⅔ CHILLED WATER COIL

AIR

⅓ CHILLED WATER COIL

CHILLED WATER SUPPLY

CHILLED WATER RETURN

**FIGURE 1-67**

CHILLED WATER SPLIT COIL PIPING DIAGRAM

NO SCALE

Figure 1-67 shows a standard arrangement of the piping of a chilled water split coil installation. This has been the source of many problems for designers. Frequently, this detail is used completely as is to avoid any misinterpretation of intent.

73

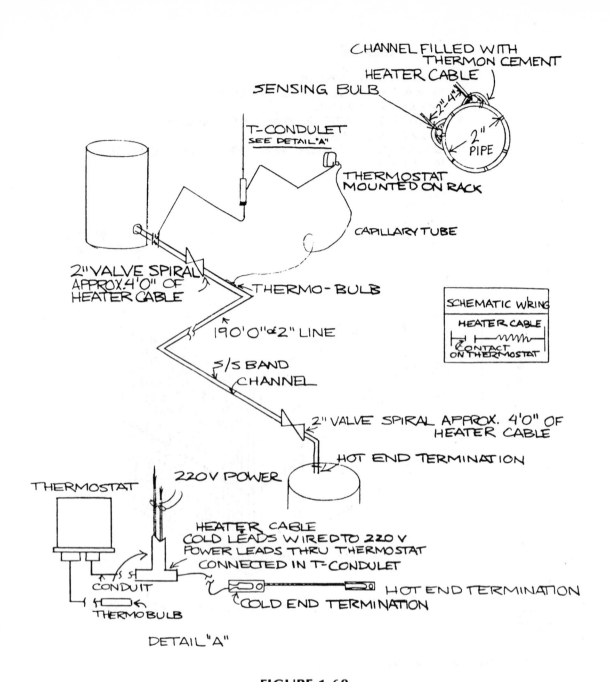

**FIGURE 1-68**

ONE PARALLEL HEATER CABLE – NO CONTACTOR
— NO SCALE —

Figure 1-68 is a detail of a typical heater cable installation on a water or other form of piping. The heater cable must go as shown from one end of the system to the other and the general relationship of sensing bulb and cable with the spacing between the two is of particular importance in this type of installation.

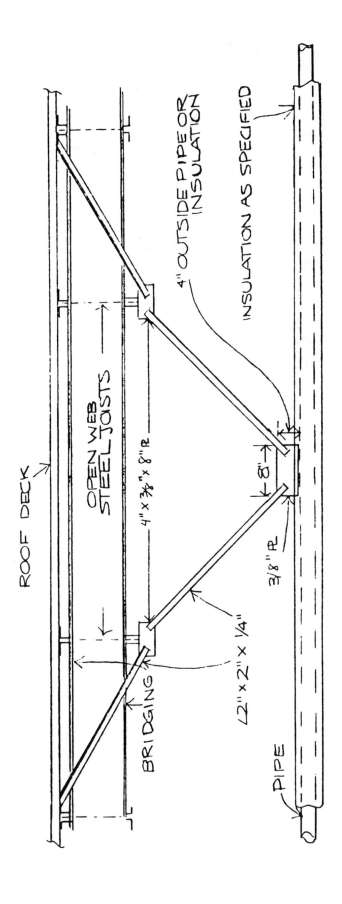

ROOF DECK

OPEN WEB STEEL JOISTS

4" × 3/8" × 8" ℞

3/8" ℞

∟ 2" × 2" × 1/4"

BRIDGING

PIPE

4" OUTSIDE PIPE OR INSULATION

INSULATION AS SPECIFIED

8"

**FIGURE 1-69**

ANCHOR WITH JOINT PERPENDICULAR TO PIPE RUN

NO SCALE

Figures 1-69 through 1-76 are a series of details showing the typical connections of anchors and pipe supports in some of the more common types of installations and, of course, they must be sized in each case to suit the pipe involved or the clearance from ground, ceiling, wall, or whatever the situation requires. In addition, the detail of an expansion joint and bolt in an old brick or concrete type of case is shown as well as a typical vibration eliminator which serves as a basic set of details commonly used in mechanical piping work.

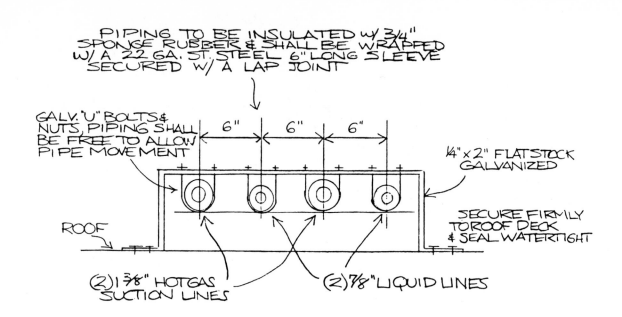

PIPING TO BE INSULATED w/ 3/4"
SPONGE RUBBER & SHALL BE WRAPPED
w/ A 22 GA. ST. STEEL 6" LONG SLEEVE
SECURED w/ A LAP JOINT

GALV. "U" BOLTS &
NUTS, PIPING SHALL
BE FREE TO ALLOW
PIPE MOVEMENT

6"    6"    6"

1/4" x 2" FLATSTOCK
GALVANIZED

ROOF

SECURE FIRMLY
TO ROOF DECK
& SEAL WATERTIGHT

(2) 1 3/8" HOT GAS
SUCTION LINES

(2) 7/8" LIQUID LINES

**FIGURE 1-70**

SUPPORTING PIPES OFF OF ROOF
NO SCALE

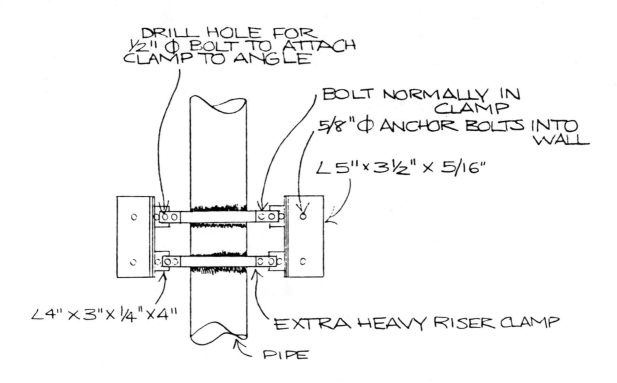

DRILL HOLE FOR
½" Ø BOLT TO ATTACH
CLAMP TO ANGLE

BOLT NORMALLY IN
CLAMP
5/8" Ø ANCHOR BOLTS INTO
WALL

∠5" × 3½" × 5/16"

∠4" × 3" × ¼" × 4"

EXTRA HEAVY RISER CLAMP

PIPE

ELEVATION

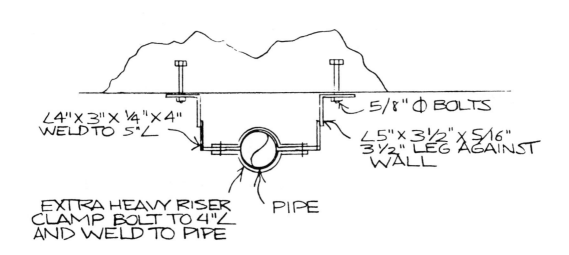

∠4" × 3" × ¼" × 4"
WELD TO 5"∠

5/8" Ø BOLTS

∠5" × 3½" × 5/16"
3½" LEG AGAINST
WALL

EXTRA HEAVY RISER
CLAMP BOLT TO 4"∠
AND WELD TO PIPE

PIPE

PLAN

**FIGURE 1-71**

ANCHOR TO VERTICAL WALL
NO SCALE

77

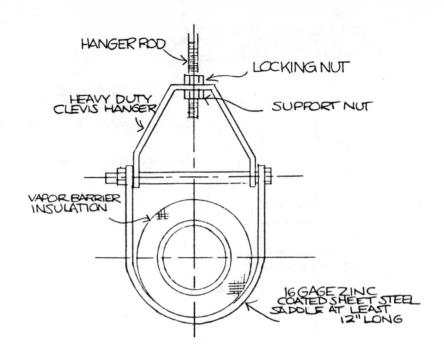

HANGER ROD

LOCKING NUT

HEAVY DUTY
CLEVIS HANGER

SUPPORT NUT

VAPOR BARRIER
INSULATION

16 GAGE ZINC
COATED SHEET STEEL
SADDLE AT LEAST
12" LONG

**FIGURE 1-72**

CLEVIS HANGER
SINGLE HORIZONTAL RUNS W/ VAPOR BARRIER INSULATION
NO SCALE

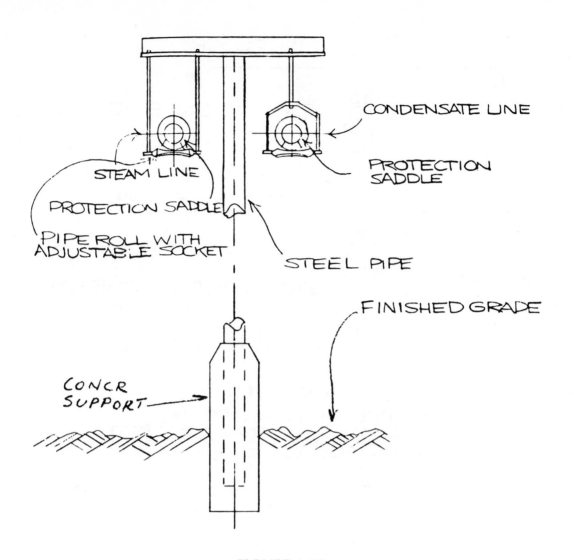

CONDENSATE LINE

PROTECTION SADDLE

STEAM LINE

PROTECTION SADDLE

PIPE ROLL WITH ADJUSTABLE SOCKET

STEEL PIPE

FINISHED GRADE

CONCR SUPPORT

**FIGURE 1-73**

TYPICAL ABOVEGROUND PIPE SUPPORTS

NO SCALE

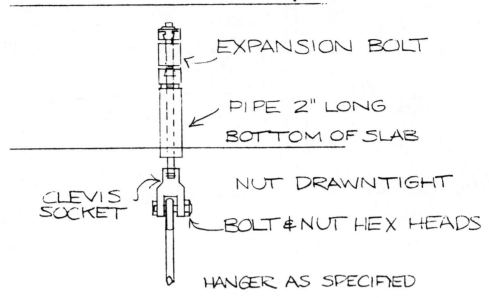

FIN. FLOOR

EXPANSION BOLT

PIPE 2" LONG

BOTTOM OF SLAB

NUT DRAWN TIGHT

CLEVIS SOCKET

BOLT & NUT HEX HEADS

HANGER AS SPECIFIED

**FIGURE 1-74**

SUPPORT FOR USE IN
OLD BRICKWORK AND CONCRETE
——— NO SCALE ———

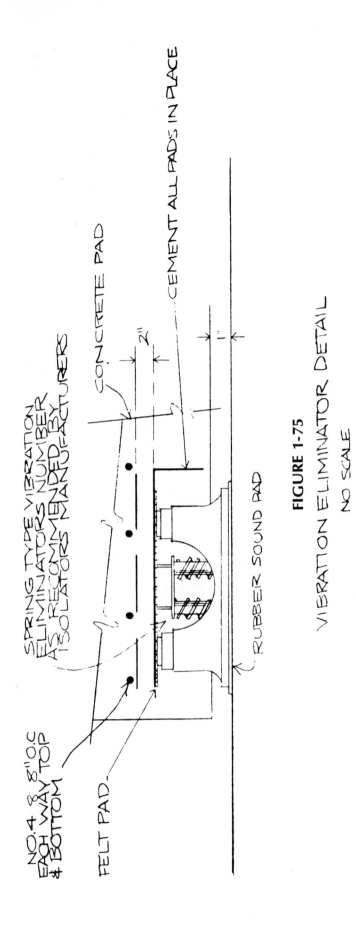

**FIGURE 1-75**

VIBRATION ELIMINATOR DETAIL

NO SCALE

82

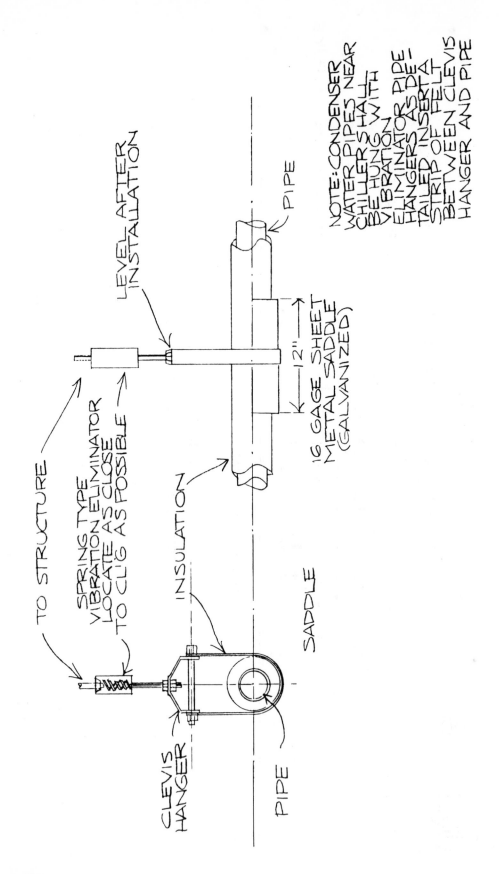

TO STRUCTURE

SPRING TYPE
VIBRATION ELIMINATOR
LOCATE AS CLOSE
TO CL'G AS POSSIBLE

LEVEL AFTER
INSTALLATION

PIPE

12"

16 GAGE SHEET
METAL SADDLE
(GALVANIZED)

NOTE: CONDENSER
WATER PIPES NEAR
CHILLERS SHALL
BE HUNG WITH
VIBRATION
ELIMINATOR PIPE
HANGERS AS DE-
TAILED INSERTA
STRIP OF FELT
BETWEEN CLEVIS
HANGER AND PIPE

INSULATION

SADDLE

CLEVIS
HANGER

PIPE

**FIGURE 1-76**

CHILLED WATER PIPE HANGERS

NO SCALE

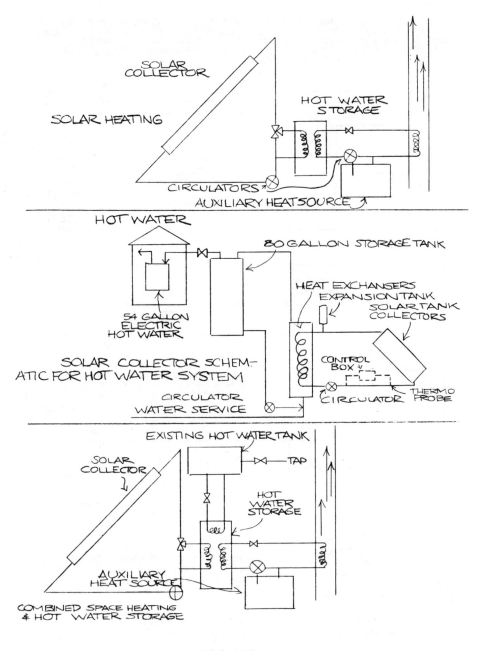

SOLAR COLLECTOR

SOLAR HEATING

HOT WATER STORAGE

CIRCULATORS

AUXILIARY HEAT SOURCE

HOT WATER

80 GALLON STORAGE TANK

54 GALLON ELECTRIC HOT WATER

HEAT EXCHANGERS
EXPANSION TANK
SOLAR TANK
COLLECTORS

SOLAR COLLECTOR SCHEMATIC FOR HOT WATER SYSTEM

CIRCULATOR
WATER SERVICE

CONTROL BOX

CIRCULATOR

THERMO PROBE

EXISTING HOT WATER TANK

SOLAR COLLECTOR

TAP

HOT WATER STORAGE

AUXILIARY HEAT SOURCE

COMBINED SPACE HEATING & HOT WATER STORAGE

## FIGURE 1-77

SOLAR HEAT SYSTEM APPLICATIONS
NO SCALE

Figures 1-77, 1-78, and 1-79 are schematic details of typical solar piping used strictly for heating, hot water heating, heating air conditioning and hot water. The individual connections to each piece of equipment are not detailed, but the general schematic arrangement is detailed and it is suggested that these details be used as the basis for a schematic arrangement and further detailing of each individual piece. The detailing of valves, unions, and so forth for each individual piece is a relatively standard type of installation.

AQUASTATS ARE REVERSE-ACTING—OPERATE TO
"CLOSE" CIRCUIT TO CIRCULATING PUMP NO 2 ON
WATER TEMPERATURE RISE IN SOLAR HEATER
AND "BREAK" CIRCUIT ON TEMPERATURE DROP
SET AT 180

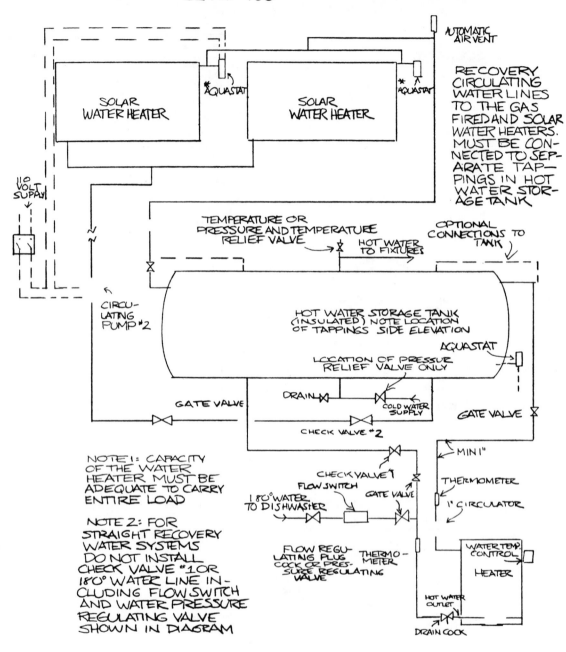

**FIGURE 1-78**

BOOSTER-RECOVERY SYSTEM USING ONE OR MORE SOLAR
WATER HEATERS AS A FUEL ECONOMIZER UTILIZING
SOLAR HEAT WHEN AVAILABLE
——————— NO SCALE ———————

84

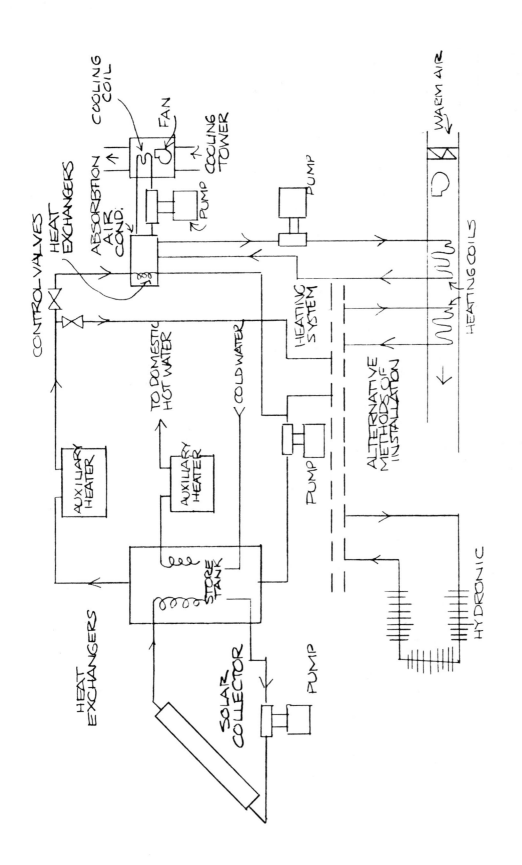

**FIGURE 1-79**

PIPING SYSTEM FOR HEATING, AIR CONDITIONING & DOMESTIC WATER
NO SCALE

85

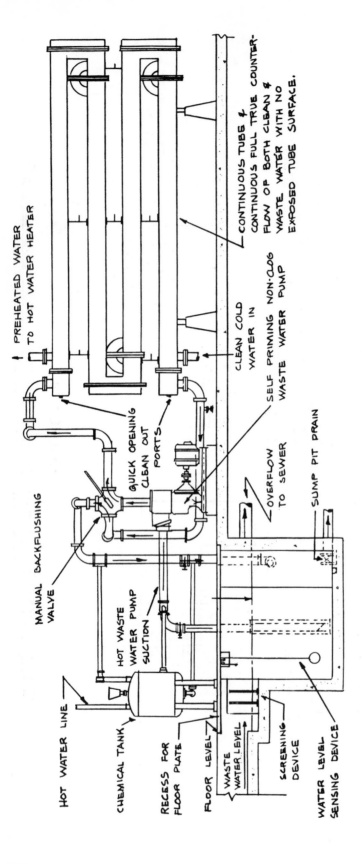

**FIGURE 1-80**

HEAT RECLAIMER SYSTEM

FOR TEXTILE MILLS, LAUNDRIES, OR WHEREVER HOT WATER IS DUMPED "DOWN THE SEWER".

Figure 1-80 indicates a standard arrangement of heat reclaiming for any installation where hot water is being wasted because of the requirements of the process. This detail is fairly standard and is commonly used in any form of heat recovery for this type of application.

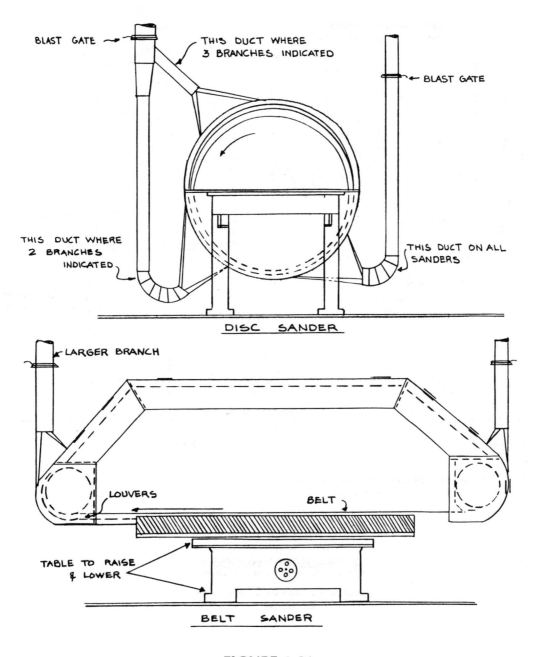

BLAST GATE

THIS DUCT WHERE
3 BRANCHES INDICATED

BLAST GATE

THIS DUCT WHERE
2 BRANCHES
INDICATED

THIS DUCT ON ALL
SANDERS

DISC SANDER

LARGER BRANCH

LOUVERS

BELT

TABLE TO RAISE
& LOWER

BELT SANDER

# FIGURE 1-81

## EXHAUST HOODS

RECOMMENDED DESIGNS FOR VARIOUS EQUIPMENT

Figures 1-81 through 1-85 are details which are typical standards for
exhaust hoods for various types of industrial equipment. There is not
much in the way of materials shown. More notation of the arrangement
and the velocities are required related to the particular project. Generally,
this type of work is done with 18 gauge welded black iron pipe, although
other materials and weights may be used depending on the particular
requirements of the design.

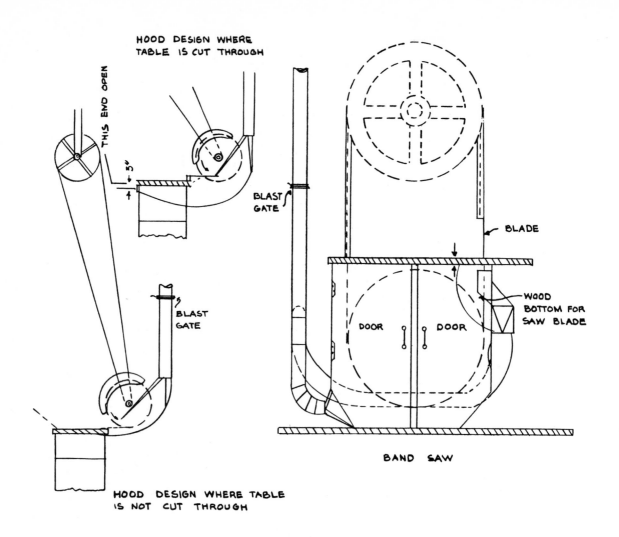

HOOD DESIGN WHERE
TABLE IS CUT THROUGH

THIS END OPEN

5"

BLAST GATE

BLAST GATE

HOOD DESIGN WHERE TABLE
IS NOT CUT THROUGH

BLADE

WOOD BOTTOM FOR SAW BLADE

DOOR    DOOR

BAND SAW

**FIGURE 1-82**

EXHAUST   HOODS

RECOMMENDED   DESIGNS   FOR   VARIOUS
EQUIPMENT

88

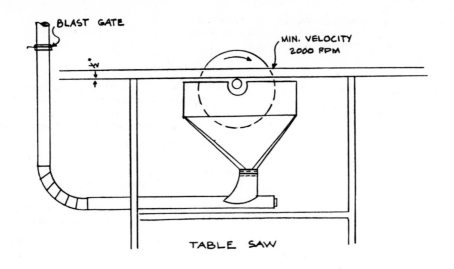

BLAST GATE

MIN. VELOCITY 2000 RPM

W

TABLE SAW

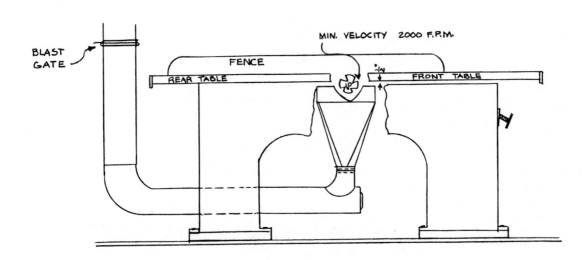

BLAST GATE

MIN. VELOCITY 2000 F.P.M.

FENCE

REAR TABLE

W

FRONT TABLE

## FIGURE 1-83

## EXHAUST HOODS

RECOMMENDED DESIGNS FOR VARIOUS EQUIPMENT

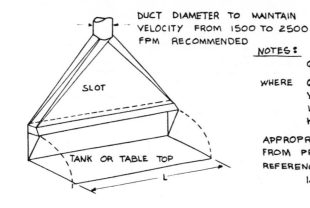

DUCT DIAMETER TO MAINTAIN VELOCITY FROM 1500 TO 2500 FPM RECOMMENDED

SLOT

TANK OR TABLE TOP

L

NOTES:

$$Q = W \times L \times K$$

WHERE  Q = AIR QUANTITY REQUIRED, CFM
W = TANK OR TABLE WIDTH, FEET
L = TANK OR TABLE LENGTH, FEET
K = VENTILATION RATE, CFM PER SQ. FT.

APPROPRIATE VALUES OF K CAN BE DETERMINED FROM PROCEDURES OUTLINED IN THE FOLLOWING REFERENCES:

1. INDUSTRIAL CODE RULES 12 AND 18 FOR OPEN SURFACE TANK OPERATIONS; DIV. OF INDUSTRIAL HYGIENE, NEW YORK STATE DEPARTMENT OF LABOR.

2. A.S.A. STANDARD Z9.1," VENTILATION AND OPERATION OF OPEN SURFACE TANKS."

3. "INDUSTRIAL VENTILATION", AMERICAN CONFERENCE OF GOVERNMENTAL INDUSTRIAL HYGIENISTS.

NOTE:

MAXIMUM WIDTH  W= 4 FEET
MINIMUM RATIO, LENGTH TO WIDTH
    L/W, 2 TO 1
MAXIMUM LENGTH OF TRANSITION
    PIECE, 4 FT
[USE MORE THAN ONE FOR
    LARGER HOODS

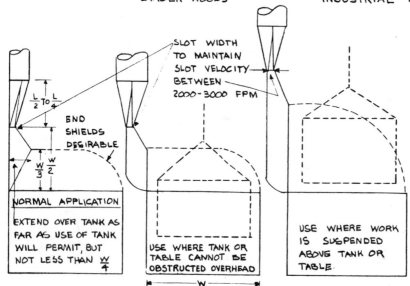

SLOT WIDTH TO MAINTAIN SLOT VELOCITY BETWEEN 2000-3000 FPM

$\frac{L}{2}$ TO $\frac{L}{4}$

END SHIELDS DESIRABLE

$\frac{W}{3}$ $\frac{W}{2}$

NORMAL APPLICATION

EXTEND OVER TANK AS FAR AS USE OF TANK WILL PERMIT, BUT NOT LESS THAN $\frac{W}{4}$

USE WHERE TANK OR TABLE CANNOT BE OBSTRUCTED OVERHEAD

W

USE WHERE WORK IS SUSPENDED ABOVE TANK OR TABLE.

**FIGURE 1-84**

REAR HOOD FOR OPEN SURFACE TANKS

—— NO SCALE ——

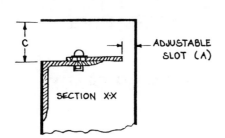

SECTION X-X

ADJUSTABLE SLOT (A)

NOTES:

1. SLOT VELOCITIES:
   SLOT A. 1,000 TO 2,000 FPM.
   SLOT C. 500 TO 1,000 FPM.

2. B & D MANIFOLD DIMENSIONS -AS LARGE AS POSSIBLE TO GIVE MINIMUM VELOCITY.

3. F AT LEAST 2½ B

4. G PIPE DIAMETER TO MAINTAIN PIPE VELOCITY BETWEEN 1,500 & 2,500 FPM.

5. GAS HEATED TANKS SHALL HAVE COMBUSTION CHAMBER INDEPENDENTLY VENTILATED TO OUTSIDE.

6. SINGLE SLOT ON LONG SIDE IS PERMITTED FOR TANKS 18" WIDE & LESS.

7. USE SLOTS ALONG BOOTH-LONG SIDES FOR TANKS WIDER THAN 18".

8. INTERNAL BAFFLES RECCOMEND FOR TANKS OVER 6 FT. IN LENGTH.

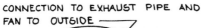

CONNECTION TO EXHAUST PIPE AND FAN TO OUTSIDE

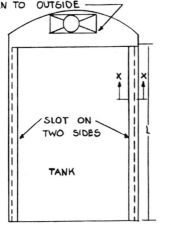

SLOT ON TWO SIDES

TANK

END SHIELDS DESIRABLE    TAPERED CONNECTION FOR LOW RESISTANCE

IF INTERNAL BAFFLES ARE USED IN MANIFOLD B, DESIGN AS SHOWN

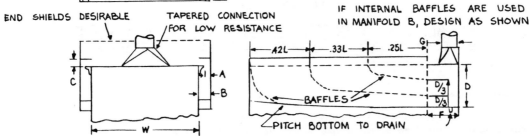

BAFFLES

PITCH BOTTOM TO DRAIN

FIGURE 1-85

LATERAL HOOD FOR OPEN SURACE TANKS

NO SCALE

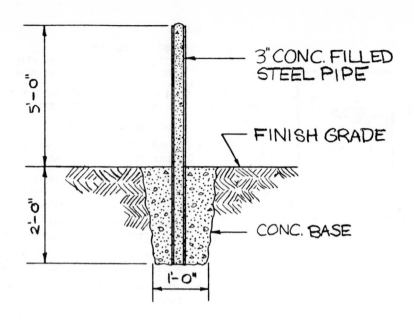

5'-0"

2'-0"

3" CONC. FILLED
STEEL PIPE

FINISH GRADE

CONC. BASE

1'-0"

GUARD POST

**FIGURE 1-86**

TYPICAL PROTECTIVE VEHICLE BUMPER
FOR EXTERIOR SURFACE MOUNTED
EQUIPMENT.

Figure 1-86 is another example of a simple detail. It is the type of bumper
that would be used singularly or in multiple around any sort of outdoor
equipment installation where protection is required from traffic parked in
the vicinity.

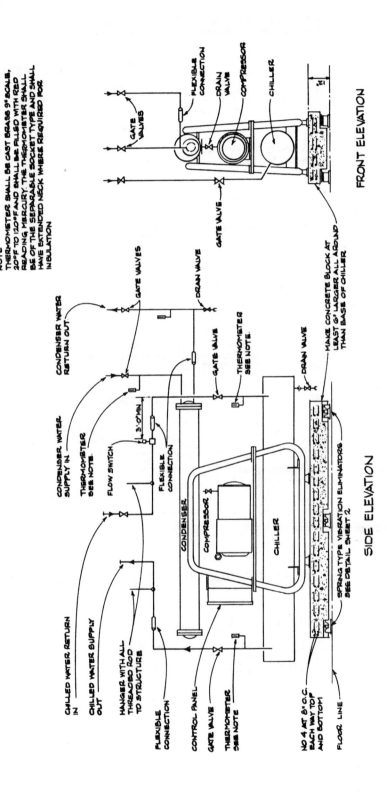

**NOTE:**
THERMOMETER SHALL BE CAST BRASS 9" SCALE, 20°F TO 120°F AND SHALL BE FILLED WITH RED READING MERCURY. THE THERMOMETER SHALL BE OF THE SEPARABLE SOCKET TYPE AND SHALL HAVE EXTENDED NECK WHERE REQUIRED FOR INSULATION.

GATE VALVES

FLEXIBLE CONNECTION
DRAIN VALVE
COMPRESSOR
CHILLER

GATE VALVE

**FRONT ELEVATION**

GATE VALVES
CONDENSER WATER RETURN OUT
DRAIN VALVE
GATE VALVE
THERMOMETER SEE NOTE.
CONDENSER WATER SUPPLY IN
THERMOMETER SEE NOTE.
FLOW SWITCH
3'-0" MIN.
FLEXIBLE CONNECTION
DRAIN VALVE
MAKE CONCRETE BLOCK AT LEAST 6" LARGER ALL AROUND THAN BASE OF CHILLER

CHILLED WATER RETURN IN
CHILLED WATER SUPPLY OUT
HANGER WITH ALL THREADED ROD TO STRUCTURE

CONDENSER
COMPRESSOR
CHILLER

SPRING TYPE VIBRATION ELIMINATORS SEE DETAIL SHEET 2

FLEXIBLE CONNECTION
CONTROL PANEL
GATE VALVE
THERMOMETER SEE NOTE

NO. 4 AT 8" O.C. EACH WAY TOP AND BOTTOM
FLOOR LINE

**SIDE ELEVATION**

## FIGURE 1-87

RECIPROCATING CHILLER INSTALLATION DETAILS
SHEET 1 OF 2

Figures 1-87 and 1-88 indicate details of the standard reciprocating chiller installation. Figure 1-88 should always accompany Figure 1-87 whenever Figure 1-87 is used. Figure 1-88 emphasizes and expands upon the troublesome parts of the detail shown in general form in Figure 1-87.

93

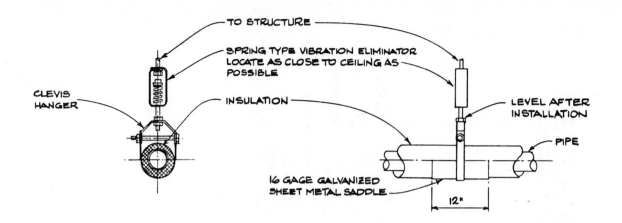

TO STRUCTURE

SPRING TYPE VIBRATION ELIMINATOR
LOCATE AS CLOSE TO CEILING AS
POSSIBLE

CLEVIS
HANGER

INSULATION

LEVEL AFTER
INSTALLATION

PIPE

16 GAGE GALVANIZED
SHEET METAL SADDLE

12"

CHILLED WATER PIPE HANGERS

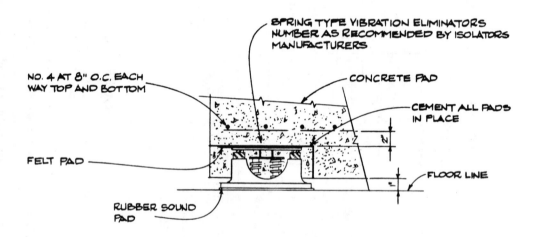

SPRING TYPE VIBRATION ELIMINATORS
NUMBER AS RECOMMENDED BY ISOLATORS
MANUFACTURERS

NO. 4 AT 8" O.C. EACH
WAY TOP AND BOTTOM

CONCRETE PAD

CEMENT ALL PADS
IN PLACE

FELT PAD

FLOOR LINE

RUBBER SOUND
PAD

VIBRATION ELIMINATOR DETAIL

**FIGURE 1-88**

RECIPROCATING CHILLER INSTALLATION DETAILS
SHEET 2 OF 2

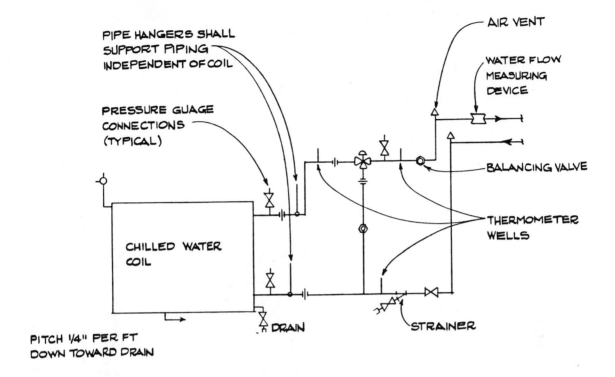

PIPE HANGERS SHALL
SUPPORT PIPING
INDEPENDENT OF COIL

AIR VENT

WATER FLOW
MEASURING
DEVICE

PRESSURE GUAGE
CONNECTIONS
(TYPICAL)

BALANCING VALVE

THERMOMETER
WELLS

CHILLED WATER
COIL

DRAIN

STRAINER

PITCH ¼" PER FT
DOWN TOWARD DRAIN

**FIGURE 1-89**

CHILLED WATER COIL PIPING

Figure 1-89 should be used as a detail to be further refined and as a
reminder to the designer that if this detail is not included in any of his ⅜
or larger plans, all of the work shown on the detail must be incorporated
on the plan.

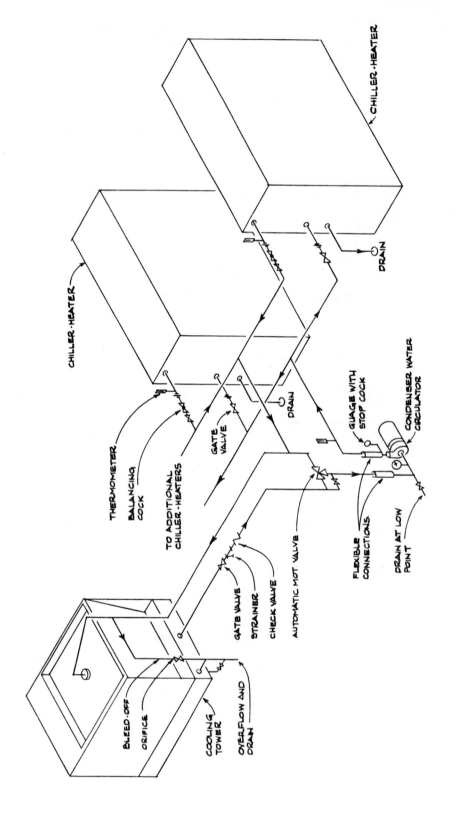

**FIGURE 1-90**

MULTIPLE CHILLER-HEATER WITH MANUAL CHANGEOVER BLENDING VALVE
(TOWER BASIN ABOVE CHILLER HEATER)

Figure 1-90 is a typical installation of a multiple chiller-heater with manual changeover blending valves. It is important to note the locations of balancing and gate valves and overflow and drain valves in the drawing.

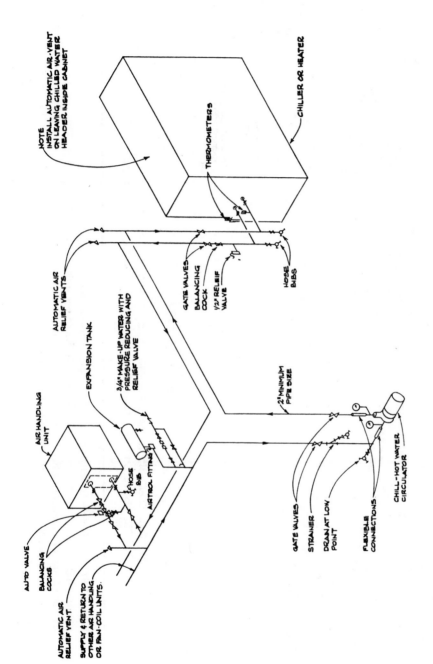

**FIGURE 1-91**

TYPICAL CHILLED-HOT WATER PIPING DIAGRAM

Figure 1-91 is a typical chilled hot water piping diagram. The expansion tank location on the suction side of the pump, as well as the 2" minimum pipe size on the discharge of the circulator to the chiller or heater, are important points in this detail.

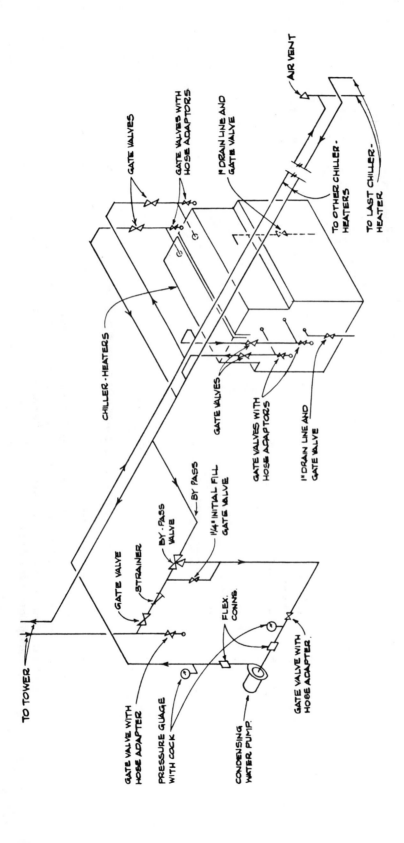

**FIGURE 1-92**

TYPICAL CONDENSING WATER PIPING DIAGRAM FOR MULTIPLE CHILLER-HEATERS
(TOWER ABOVE UNITS-MANUAL CHANGEOVER OF CONDENSING WATER CIRCUIT)

Figure 1-92 is fairly self-explanatory. It is the typical arrangement of condensing water piping for multiple chillers with the tower units above and a manual changeover of the condensing water circuit. This detail requires additional refinement in fittings and valves to become complete in every aspect.

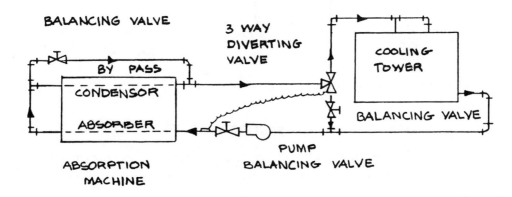

**FIGURE 1-93**

## CONDENSOR WATER TEMPERATURE CONTROL

Figure 1-93 illustrates the arrangement between the condenser and the
cooling tower to control water temperature. Note that the three way valve
is in a different position when the cooling tower elevation is not sufficient
to overcome the losses in the valve and piping.

99

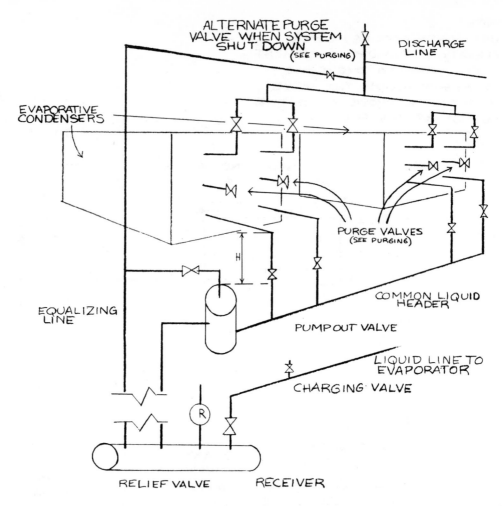

ALTERNATE WITH COMMON LIQUID HEADER ABOVE RECEIVER LEVEL

**FIGURE 1-94**

PARALLELING EVAPORATIVE CONDENSERS WITH ONE OR MORE COMPRESSORS
———————————— NO SCALE ————————————

Figure 1-94 is a sketch type drawing used to clarify the work to be shown
on the plan. It is a typical basic schematic used in preliminary planning.

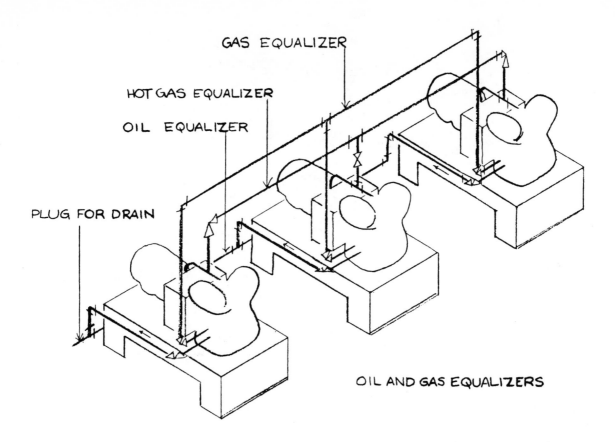

GAS EQUALIZER

HOT GAS EQUALIZER

OIL EQUALIZER

PLUG FOR DRAIN

OIL AND GAS EQUALIZERS

**FIGURE 1-95**

TYPICAL INTERCONNECTING PIPING
FOR MULTIPLE CONDENSERS

*NO SCALE*

Figure 1-95 is a drawing to indicate equalizing lines. An interconnecting
oil equalizing line is needed between all crankcases. The oil equalizer
may be run level or it may be run at the floor. Under no condition should
it be run higher than that of the compressor tappings. Gas equalization
lines may be run at the same height as the tappings or above them.

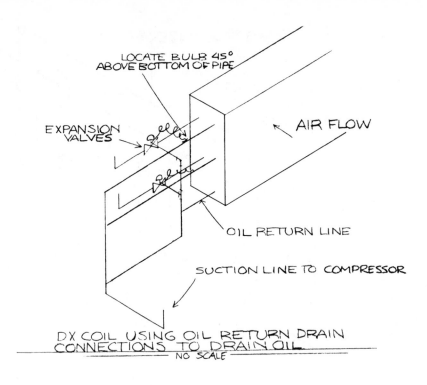

LOCATE BULB 45°
ABOVE BOTTOM OF PIPE

AIR FLOW

EXPANSION
VALVES

OIL RETURN LINE

SUCTION LINE TO COMPRESSOR

DX COIL USING OIL RETURN DRAIN
CONNECTIONS TO DRAIN OIL
NO SCALE

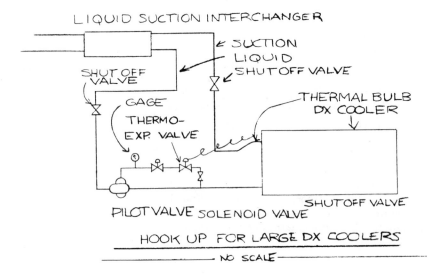

LIQUID SUCTION INTERCHANGER

SUCTION

SHUT OFF
VALVE

LIQUID
SHUTOFF VALVE

GAGE

THERMAL BULB
DX COOLER

THERMO-
EXP. VALVE

SHUTOFF VALVE

PILOT VALVE SOLENOID VALVE

HOOK UP FOR LARGE DX COOLERS

NO SCALE

**FIGURE 1-96**

Figure 1-96 shows the DX cooler which illustrates the need for a continuous liquid bleed line from some point below the liquid level in the shell in the suction line to assure the required return of oil to the compressor. Normally this is drained into the suction line. The interchanger is required to evaporate any liquid refrigerant from the refrigerant oil mixture which is continuously fed into the suction line. The DX coil shown at the top of the sheet again illustrates the use of the oil return line.

102

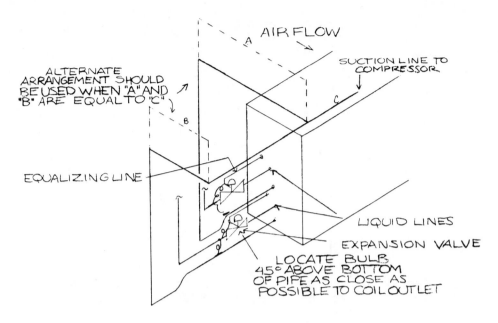

ALTERNATE ARRANGEMENT SHOULD BE USED WHEN "A" AND "B" ARE EQUAL TO "C"

AIR FLOW

SUCTION LINE TO COMPRESSOR

EQUALIZING LINE

LIQUID LINES

EXPANSION VALVE

LOCATE BULB 45° ABOVE BOTTOM OF PIPE AS CLOSE AS POSSIBLE TO COIL OUTLET

DX COIL USING SUCTION
CONNECTIONS TO DRAIN COIL
SUCTION HEADER ABOVE COIL
NO SCALE

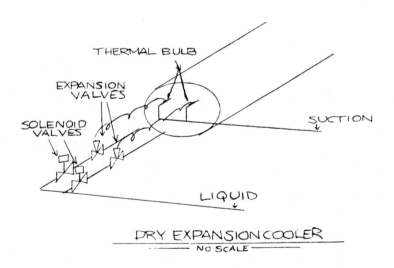

THERMAL BULB

EXPANSION VALVES

SOLENOID VALVES

SUCTION

LIQUID

DRY EXPANSION COOLER
NO SCALE

**FIGURE 1-97**

Figure 1-97 is fairly common in its application. Particular attention should be paid to the alternate arrangement noted in the upper level drawing and in the location of expansion valve sensing elements.

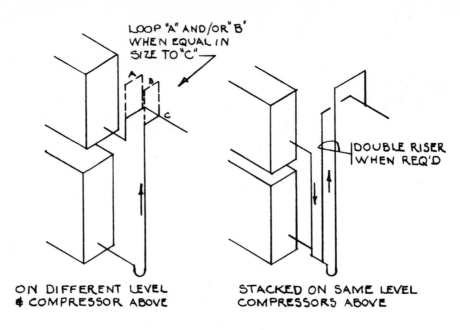

LOOP "A" AND/OR "B" WHEN EQUAL IN SIZE TO "C"

DOUBLE RISER WHEN REQ'D

ON DIFFERENT LEVEL & COMPRESSOR ABOVE

STACKED ON SAME LEVEL COMPRESSORS ABOVE

## MULTIPLE EVAPORATORS

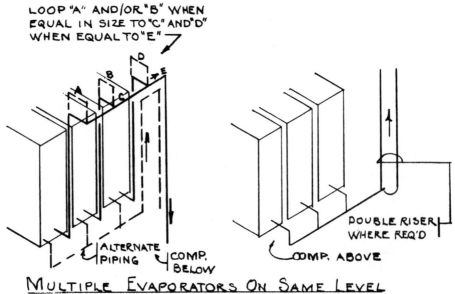

LOOP "A" AND/OR "B" WHEN EQUAL IN SIZE TO "C" AND "D" WHEN EQUAL TO "E"

ALTERNATE PIPING

COMP. BELOW

DOUBLE RISER WHERE REQ'D

COMP. ABOVE

## MULTIPLE EVAPORATORS ON SAME LEVEL

## FIGURE 1-98

## STANDARD ARRANGEMENT- SUCTION LINE LOOPS-ONE CIRC. COILS

—— NO SCALE ——

Figure 1-98 shows various typical refrigerant piping layouts and the important thing to note is the use of the inverted loop. On the drawings are dotted alternative arrangements and the entire piping system shown should be reviewed by the designer in charge and the proper one should be selected for application.

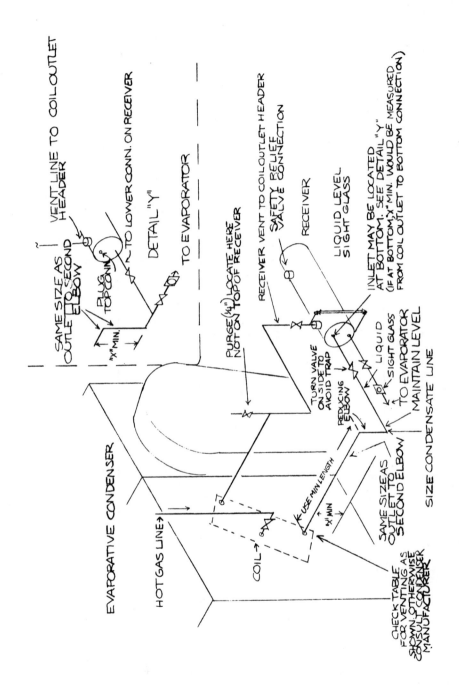

**FIGURE 1-99**

HOT GAS AND LIQUID PIPING SINGLE COIL UNIT WITH RECEIVER VENT
NO SCALE

Figure 1-99 is essentially a design guidance diagram for designers with basic information and points to carefully check before completing the design of hot gas and liquid piping as shown in this detail.

105

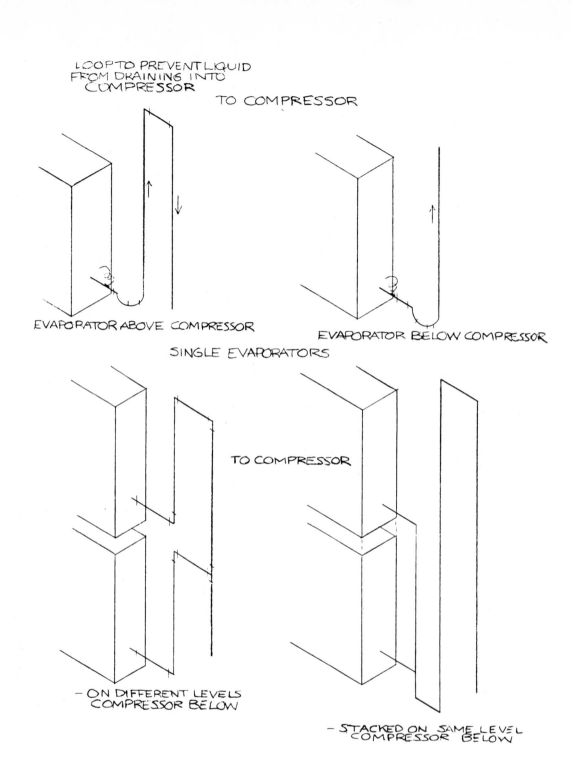

LOOP TO PREVENT LIQUID FROM DRAINING INTO COMPRESSOR

TO COMPRESSOR

EVAPORATOR ABOVE COMPRESSOR

EVAPORATOR BELOW COMPRESSOR

SINGLE EVAPORATORS

TO COMPRESSOR

- ON DIFFERENT LEVELS COMPRESSOR BELOW

- STACKED ON SAME LEVEL COMPRESSOR BELOW

**FIGURE 1-100**

STANDARD ARRANGEMENTS OF SUCTION LINE LOOPS (ONE CIRCUIT COILS SHOWN)

NO SCALE

Figure 1-100 is again a preliminary schematic that is used best as a suggestive type of design guidance drawing.

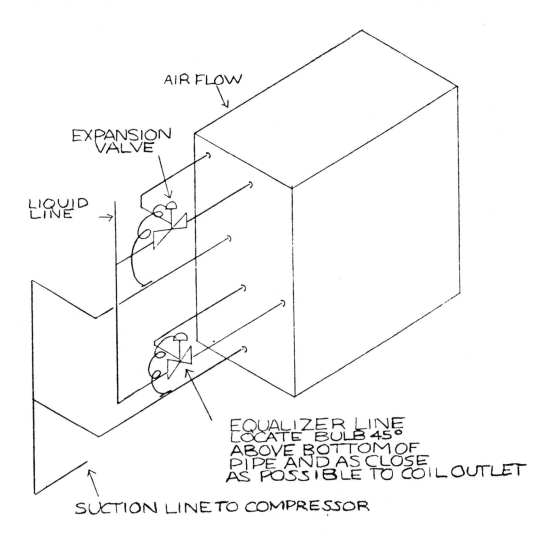

AIR FLOW

EXPANSION VALVE

LIQUID LINE

EQUALIZER LINE
LOCATE BULB 45°
ABOVE BOTTOM OF
PIPE AND AS CLOSE
AS POSSIBLE TO COIL OUTLET

SUCTION LINE TO COMPRESSOR

**FIGURE 1-101**

DX COIL USING SUCTION CONNECTIONS
TO DRAIN COIL, SUCTION HEADER
BELOW COIL
NO SCALE

Figure 1-101 shows direct expansion coil piping arrangements in which
the suction connections drain the coil headers effectively. Such an
arrangement is preferred for this type of application wherever possible.

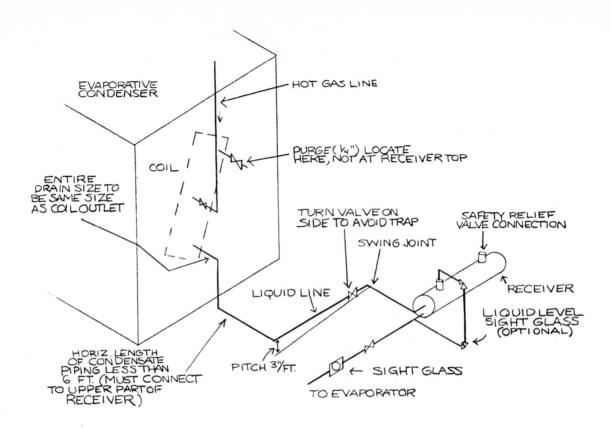

EVAPORATIVE CONDENSER

HOT GAS LINE

COIL

PURGE (¼") LOCATE HERE, NOT AT RECEIVER TOP

ENTIRE DRAIN SIZE TO BE SAME SIZE AS COIL OUTLET

TURN VALVE ON SIDE TO AVOID TRAP

SWING JOINT

SAFETY RELIEF VALVE CONNECTION

LIQUID LINE

RECEIVER

LIQUID LEVEL SIGHT GLASS (OPTIONAL)

HORIZ. LENGTH OF CONDENSATE PIPING LESS THAN 6 FT. (MUST CONNECT TO UPPER PART OF RECEIVER)

PITCH 3"/FT.

SIGHT GLASS

TO EVAPORATOR

**FIGURE 1-102**

HOT GAS AND LIQUID PIPING, SINGLE COIL UNIT WITHOUT RECEIVER VENT
NO SCALE

Figure 1-102 shows a single evaporative condenser and receiver vented back through the condensate drain line to the condensing coil outlet. Such an arrangement is applicable to a close coupled system. A separate vent is not required. It is limited to a horizontal length of condensate line of less than 6 ft. The entire condensate line from the condenser to the receiver is the same size as the coil outlet. All lines should be pitched as shown.

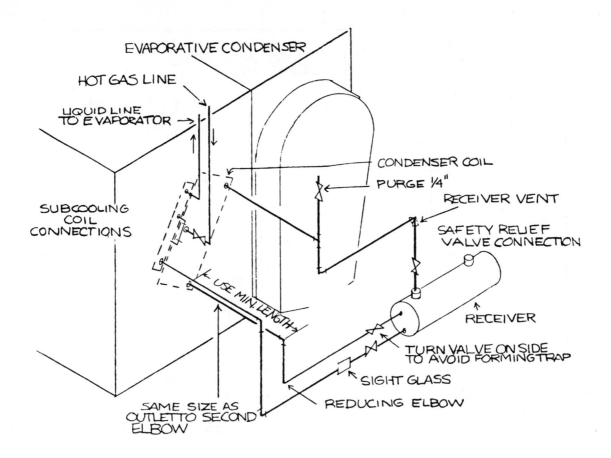

EVAPORATIVE CONDENSER

HOT GAS LINE

LIQUID LINE TO EVAPORATOR →

SUBCOOLING COIL CONNECTIONS

CONDENSER COIL

PURGE ¼"

RECEIVER VENT

SAFETY RELIEF VALVE CONNECTION

USE MIN. LENGTH

RECEIVER

TURN VALVE ON SIDE TO AVOID FORMING TRAP

SIGHT GLASS

REDUCING ELBOW

SAME SIZE AS OUTLET TO SECOND ELBOW

**FIGURE 1-103**

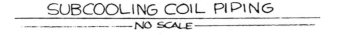

SUBCOOLING COIL PIPING

NO SCALE

Figure 1-103 shows the refrigerant piping for a single unit with a receiver vent. Note that the condensate lines from the condenser are the full size of the outlet connection and are not reduced until the second elbow is reached. This arrangement prevents trapping of liquid in the condenser coil.

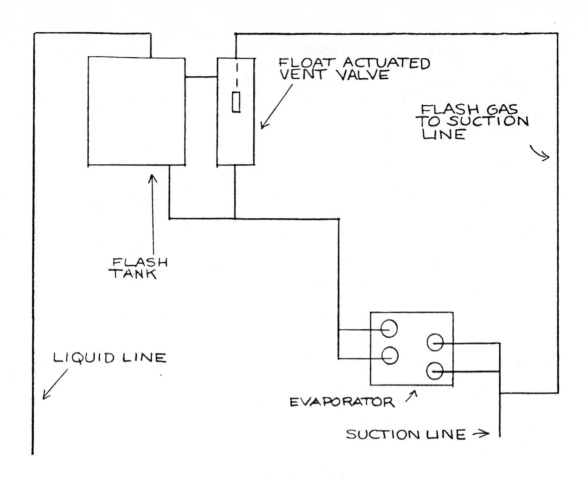

**FIGURE 1-104**

METHOD OF OVERCOMING — EFFECTS OF SYSTEM
HIGH STATIC HEAD

— NO SCALE —

Figure 1-104 shows an arrangement of a method which may be used to
overcome the effect of excessive flash gas caused by a high static head in
the system. The arrangement does not prevent the forming of flash gas,
but does offset the effect it might have on the operation of the evaporator
and the values.

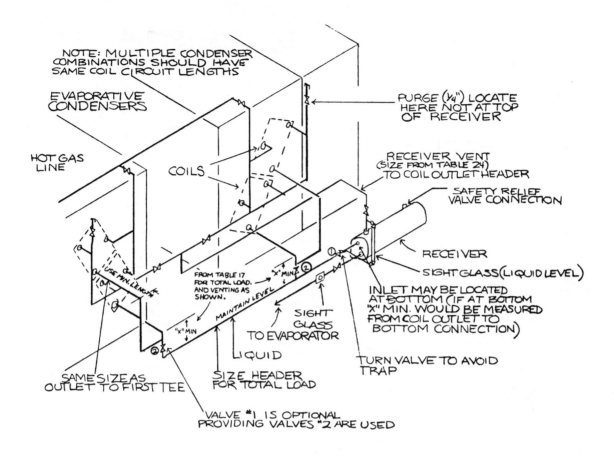

**FIGURE 1-105**

HOT GAS AND LIQUID PIPING, MULTIPLE DOUBLE COIL UNIT

NO SCALE

Figure 1-105 shows a piping arrangement for multiple units. Note the individual hot gas and vent valves for each unit. These valves prevent the operation of one unit while the other is shut down and they are essential because the idle unit at lower pressure causes hot gas to blow through the operating unit into the liquid line. Purge cocks are shown for each unit.

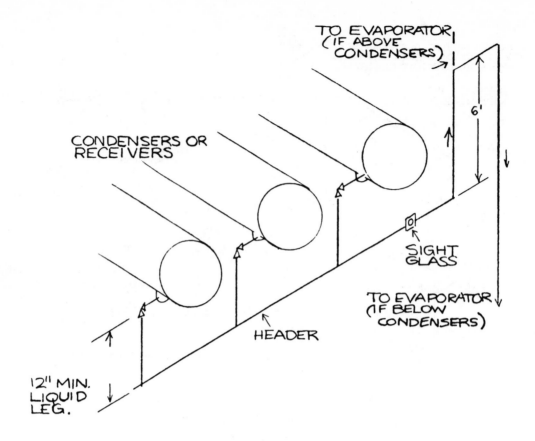

**FIGURE 1-106**

LIQUID PIPING TO INSURE CONDENSATE FLOW FROM
INTERCONNECTED CONDENSERS

NO SCALE

Figure 1-106 shows a multiple arrangement of condensers and the balancing of liquid and the 12″ required to prevent gas from blowing through. An inverted loop of at least 6 ft is recommended in the liquid line to prevent siphoning of the liquid into the evaporator during shutdown. The loop is unnecessary when liquid line solenoid valves are used.

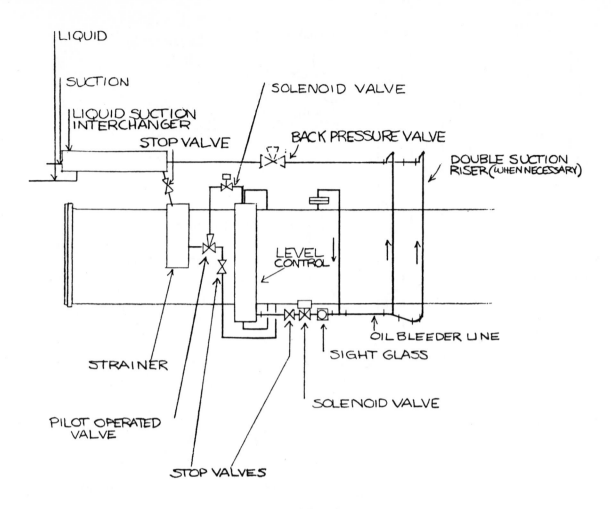

**FIGURE 1-107**

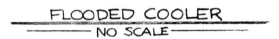

FLOODED COOLER

—— NO SCALE ——

Figure 1-107 illustrates the point of the level control continuous bleed of refrigerant liquid and the assuring of returning oil to the compressor. The drain lines should be equipped with a hand valve, solenoid valve, and sight glass. The suction interchanger is required to evaporate any liquid refrigerant oil mixtures. The water supply should never be throttled and should never bypass the cooler to avoid freeze up.

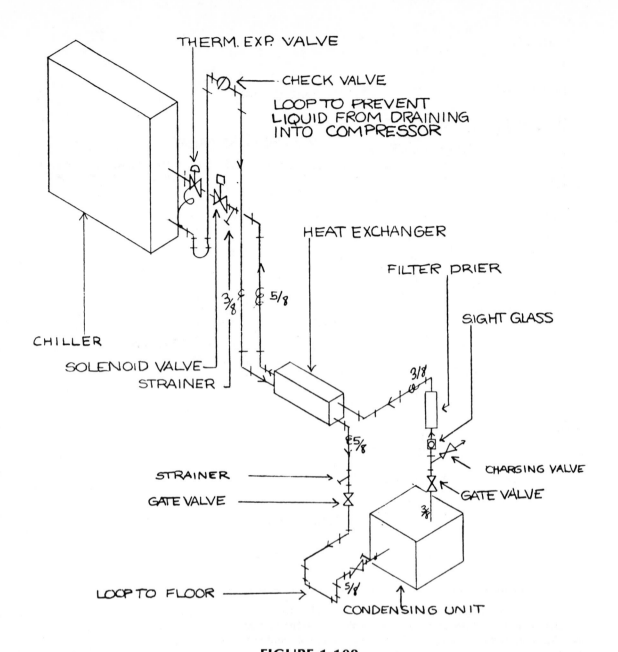

**FIGURE 1-108**

REFRIGERANT PIPING DETAIL
——— NO SCALE ———

Figure 1-108 is the typical piping of a small installation of condensing unit
and chiller and can be followed with appropriate sizing for almost any
installation of this type.

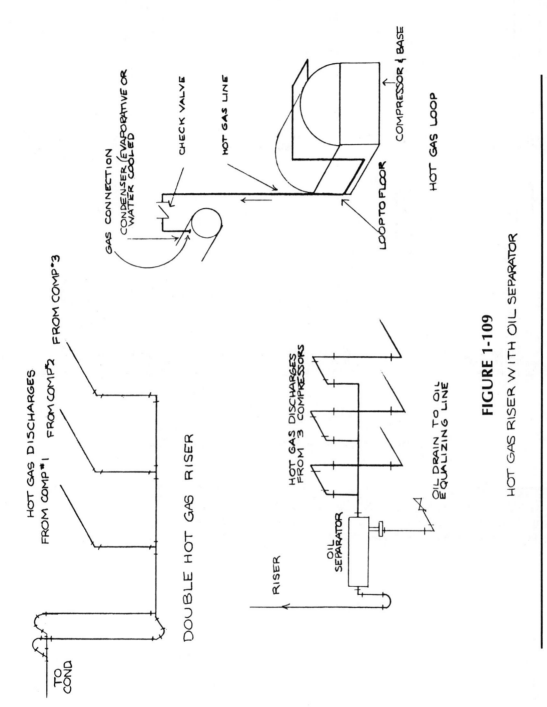

GAS CONNECTION

CONDENSER (EVAPORATIVE OR WATER COOLED)

CHECK VALVE

HOT GAS LINE

COMPRESSOR & BASE

HOT GAS LOOP

LOOP TO FLOOR

HOT GAS DISCHARGES

FROM COMP #1    FROM COMP#2    FROM COMP #3

DOUBLE HOT GAS RISER

TO COND

HOT GAS DISCHARGES FROM 3 COMPRESSORS

OIL DRAIN TO OIL EQUALIZING LINE

OIL SEPARATOR

RISER

## FIGURE 1-109

HOT GAS RISER WITH OIL SEPARATOR

Figure 1-109 is another design guidance drawing to be used to develop both the system shown on the plans and the details that will be required.

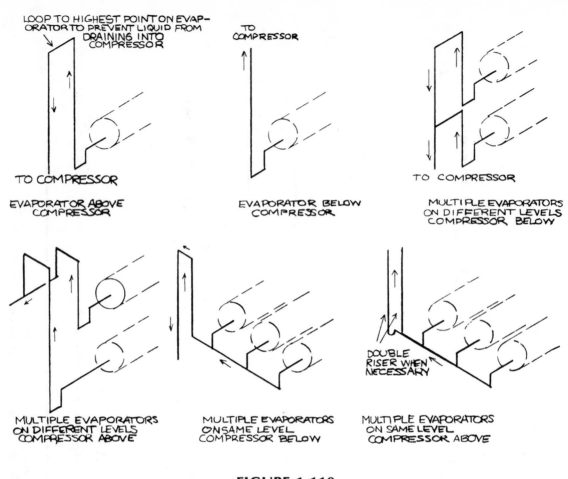

**FIGURE 1-110**

TYPICAL SUCTION LINE PIPING AT EVAPORATOR COILS

NO SCALE

Figure 1-110 is a guidance drawing so that designers and draftsmen may clearly understand the general run of piping to various types of suction connections at evaporative coils in various combinations and positions.

116

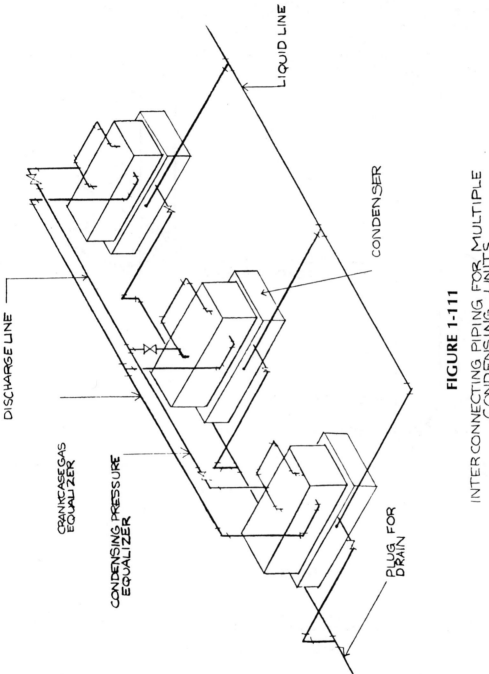

LIQUID LINE

DISCHARGE LINE

CRANKCASE GAS EQUALIZER

CONDENSING PRESSURE EQUALIZER

CONDENSER

PLUG FOR DRAIN

**FIGURE 1-111**

INTERCONNECTING PIPING FOR MULTIPLE CONDENSING UNITS
— NO SCALE —

Figure 1-111 is another preliminary type of sketch for the layout of the interconnection piping for multiple condensing units.

117

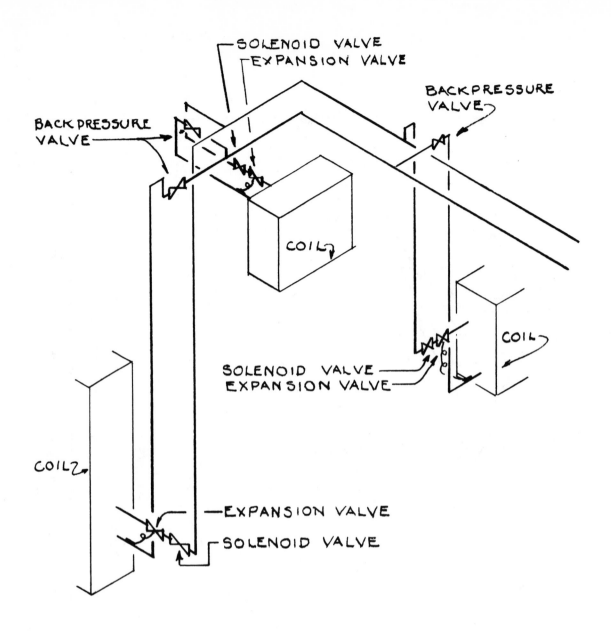

**FIGURE 1-112**

## TYPICAL INSTALLATION USING BACKPRESSURE VALVES

——— NO SCALE ———

Figure 1-112 is a relatively straightforward drawing which is used for illustrative purposes to make certain that the back pressure valves should be installed in the system and also where they should be installed.

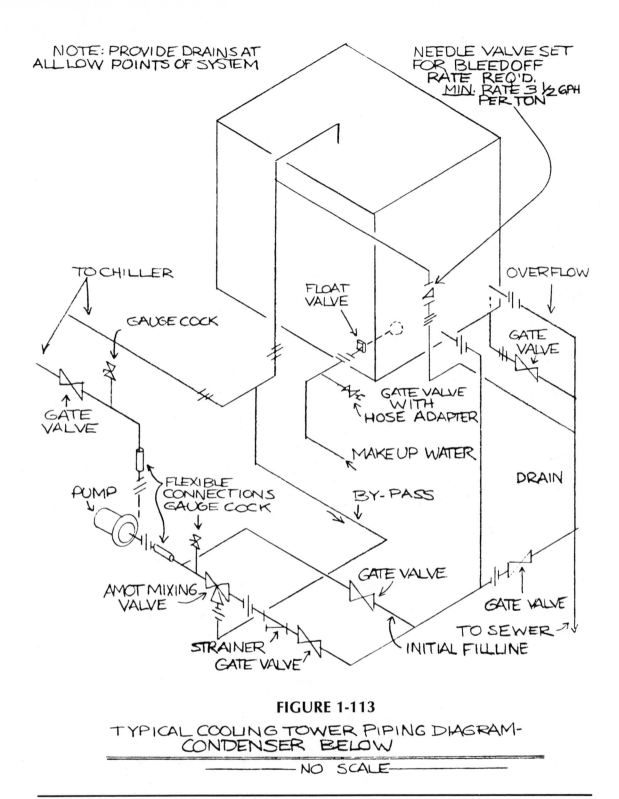

NOTE: PROVIDE DRAINS AT ALL LOW POINTS OF SYSTEM

NEEDLE VALVE SET FOR BLEEDOFF RATE REQ'D. MIN. RATE 3 ½ GPH PER TON

TO CHILLER

GAUGE COCK

OVERFLOW

FLOAT VALVE

GATE VALVE

GATE VALVE

GATE VALVE WITH HOSE ADAPTER

MAKE UP WATER

PUMP

FLEXIBLE CONNECTIONS GAUGE COCK

BY-PASS

DRAIN

AMOT MIXING VALVE

GATE VALVE

STRAINER
GATE VALVE

GATE VALVE

TO SEWER

INITIAL FILL LINE

**FIGURE 1-113**

TYPICAL COOLING TOWER PIPING DIAGRAM-
CONDENSER BELOW

———— NO SCALE ————

Figures 1-113, 1-114, and 1-115 are schematic, three-dimensional details that indicate the piping relationships for the more common type of cooling tower installation. With greater detailing and sizing of piping and valves, these drawings may be incorporated in whole or in part in a given project.

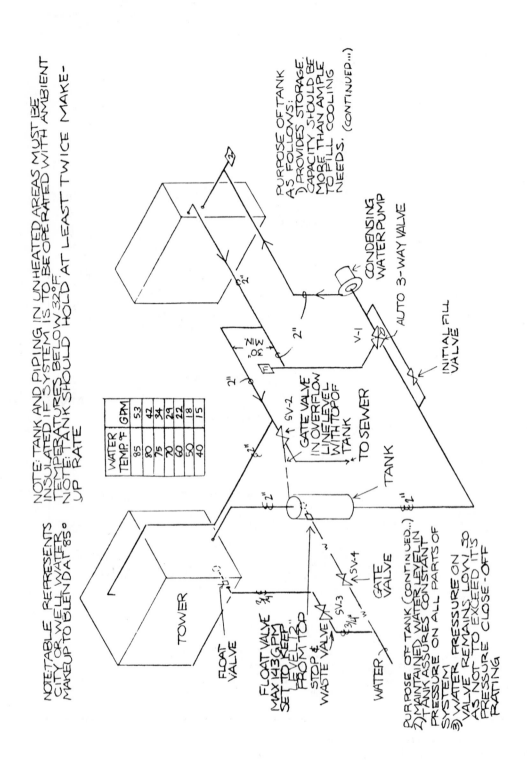

NOTE: TABLE REPRESENTS CITY OR WELL WATER MAKE-UP TO BLEND AT 85°

NOTE: TANK AND PIPING IN UNHEATED AREAS MUST BE INSULATED IF SYSTEM IS TO BE OPERATED WITH AMBIENT TEMPERATURES BELOW 32°F.
NOTE: TANK SHOULD HOLD AT LEAST TWICE MAKE-UP RATE

PURPOSE OF TANK AS FOLLOWS:
1) PROVIDES STORAGE. CAPACITY SHOULD BE MORE THAN AMPLE TO FILL COOLING NEEDS. (CONTINUED...)

| WATER TEMP.°F | GPM |
|---|---|
| 85 | 53 |
| 80 | 42 |
| 75 | 34 |
| 70 | 29 |
| 60 | 22 |
| 50 | 18 |
| 40 | 15 |

CONDENSING WATER PUMP

AUTO 3-WAY VALVE

V-1

INITIAL FILL VALVE

GATE VALVE IN OVERFLOW LINE LEVEL WITH TOP OF TANK

TO SEWER

TANK

GV-2

2"

2"

30" MIN.

2"

2"

2"

2"

TOWER

FLOAT VALVE

FLOAT VALVE MAX 143 GPM SET TO KEEP LEVEL 12" FROM TOP

STOP & WASTE VALVE

SV-3

SV-4

GATE VALVE

WATER

PURPOSE OF TANK (CONTINUED...)
2) MAINTAINED WATER LEVEL IN TANK ASSURES CONSTANT PRESSURE ON ALL PARTS OF SYSTEM
3) WATER PRESSURE ON VALVE REMAINS LOW, SO AS NOT TO EXCEED IT'S PRESSURE CLOSE-OFF RATING

**FIGURE 1-114**

MANUAL CHANGEOVER - COMBINATION TOWER
—— NO SCALE ——

120

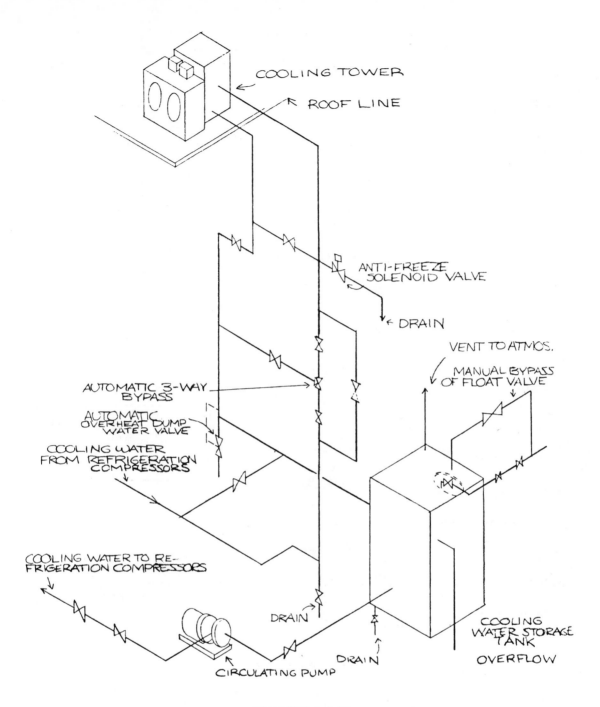

COOLING TOWER

ROOF LINE

ANTI-FREEZE SOLENOID VALVE

DRAIN

VENT TO ATMOS.

MANUAL BYPASS OF FLOAT VALVE

AUTOMATIC 3-WAY BYPASS

AUTOMATIC OVERHEAT DUMP WATER VALVE

COOLING WATER FROM REFRIGERATION COMPRESSORS

COOLING WATER TO REFRIGERATION COMPRESSORS

DRAIN

CIRCULATING PUMP

DRAIN

COOLING WATER STORAGE TANK

OVERFLOW

**FIGURE 1-115**

CONNECTIONS FOR A NON-FREEZE COOLING TOWER

NO SCALE

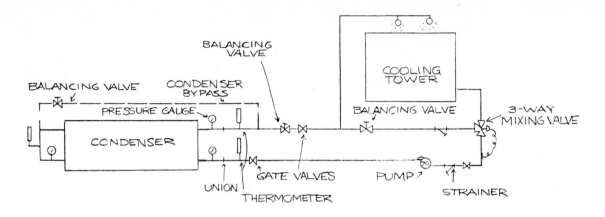

**FIGURE 1-116**

COOLING TOWER PIPING, 3 WAY MIXING VALVE
———— NO SCALE ————

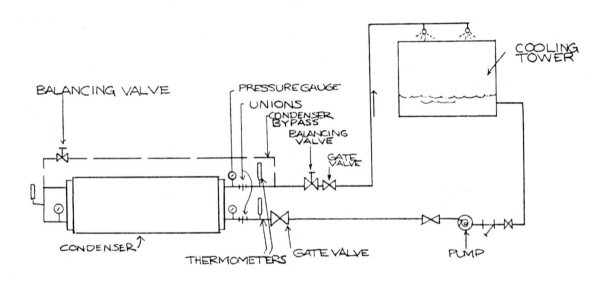

**FIGURE 1-117**

COOLING TOWER PIPING WITHOUT COOLING
WATER TEMPERATURE CONTROL
————————— NO SCALE ————

Figures 1-116 and 1-117 are schematic diagrams of specialized connections which are frequently not clearly covered. The detailing of fittings and the piping properly sized for the individual job is a design requirement.

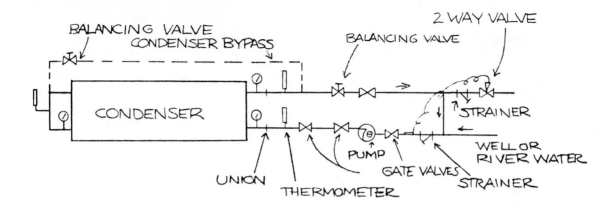

## FIGURE 1-118

PIPING FOR WELL OR RIVER WATER
NO SCALE

Figure 1-118 is the schematic piping of cooling water from a relatively uncommon source—well or river. Additional strainers and water treatment may be required in a given installation.

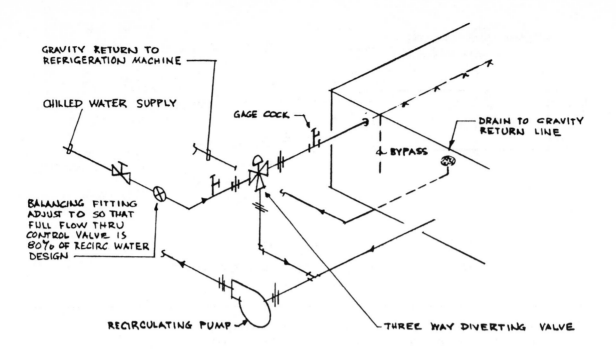

**FIGURE 1-119**

## AIR WASHER PIPING USING A THREE WAY CONTROL VALVE

Figure 1-119 is an illustration of air-washer piping with chilled water to accomplish dehumdification. The plug cock is adjusted so that the full flow through the three way diverting valve is approximately 90% of the circulating water quantity.

124

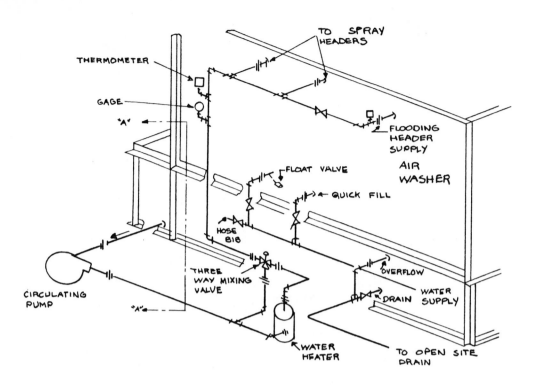

TO SPRAY HEADERS

THERMOMETER

GAGE

"A"

FLOODING HEADER SUPPLY

AIR WASHER

FLOAT VALVE

QUICK FILL

HOSE BIB

THREE WAY MIXING VALVE

CIRCULATING PUMP

"A"

WATER HEATER

OVERFLOW

WATER SUPPLY

DRAIN

TO OPEN SITE DRAIN

**FIGURE 1-120**

AIR   WASHER   PIPING   FOR   DEHUMIDIFYING   SYSTEM

NO  SCALE

Figure 1-120 is a typical layout for an air washer used for humidifying.
When the pump and the air washer are on the same level there is usually
a small suction head available for the pump. Therefore, a strainer should
be located on the discharge side of the pump. The air washer normally
has a permanent type screen at the suction to the washer to remove large
sized foreign matter. The drain line arrangement should always be
checked for compliance to local codes.

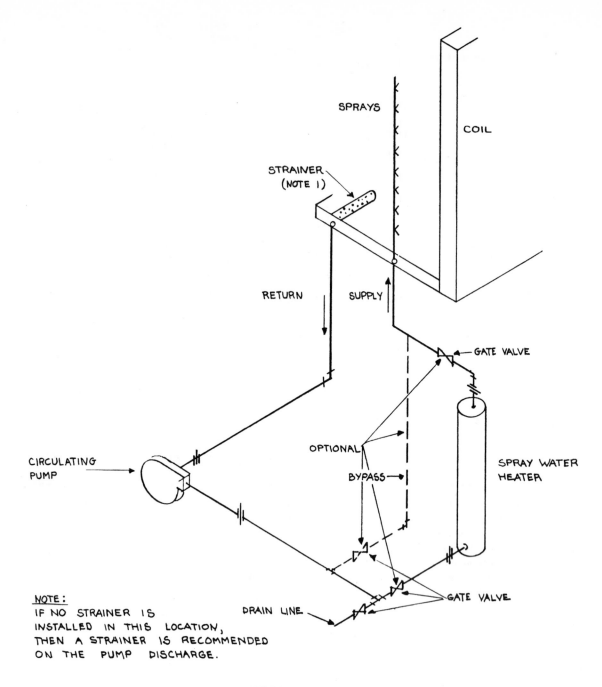

SPRAYS

COIL

STRAINER (NOTE 1)

RETURN

SUPPLY

GATE VALVE

OPTIONAL

CIRCULATING PUMP

BYPASS

SPRAY WATER HEATER

GATE VALVE

NOTE:
IF NO STRAINER IS INSTALLED IN THIS LOCATION, THEN A STRAINER IS RECOMMENDED ON THE PUMP DISCHARGE.

DRAIN LINE

## FIGURE 1-121

## SPRAY WATER COIL WITH WATER HEATER

NO SCALE

Figure 1-121 is a typical layout for a sprayed coil piping system with a water heater which may be required for humidification. If a preheat coil is used, the water heater may be eliminated. The drain lines should be fitted with a gate valve rather than a globe valve.

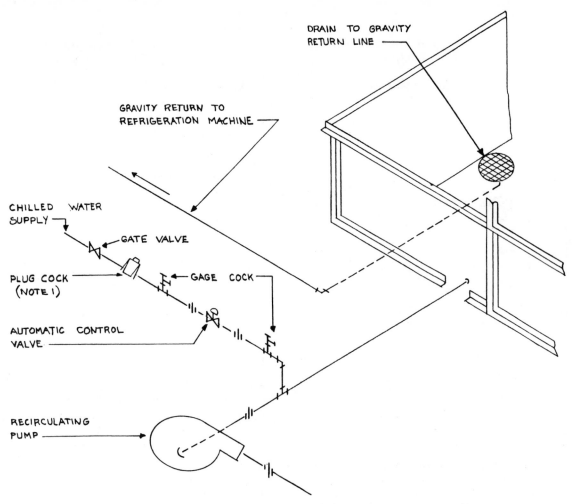

DRAIN TO GRAVITY
RETURN LINE

GRAVITY RETURN TO
REFRIGERATION MACHINE

CHILLED WATER
SUPPLY

GATE VALVE

PLUG COCK
(NOTE 1)

GAGE COCK

AUTOMATIC CONTROL
VALVE

RECIRCULATING
PUMP

NOTE: ADJUST PLUG COCK SO THAT FULL FLOW THRU DIVERTING VALVE IS
APPROXIMATELY 80% OF RECIRCULATING WATER DESIGN.

**FIGURE 1-122**

AIR WASHER PIPING USING A TWO-WAY
CONTROL VALVE

NO SCALE

Figure 1-122 is similar to Figure 1-119, except that in this case a two way
valve is used in lieu of a three way valve.

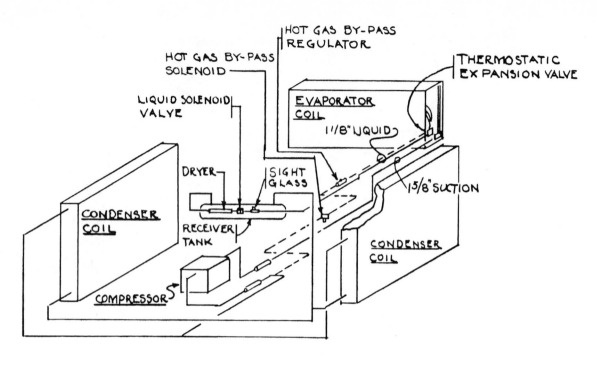

**FIGURE 1-123**

ROOF TOP UNIT—REFRIGERATION SCHEMATIC

———————— NOT TO SCALE ————————

Figure 1-123 shows the standard components in a roof-top unit. These
may all be part of a roof-top package, but there are occasions when one
or more of the items are separated from the other and the piping is
shown schematically so that the designer understands the relationship of
one part to the other.

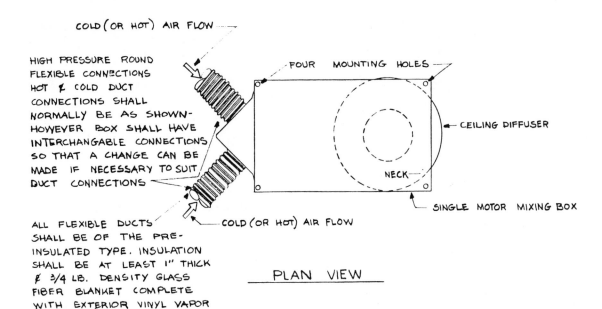

COLD (OR HOT) AIR FLOW

HIGH PRESSURE ROUND
FLEXIBLE CONNECTIONS
HOT & COLD DUCT
CONNECTIONS SHALL
NORMALLY BE AS SHOWN-
HOWEVER BOX SHALL HAVE
INTERCHANGABLE CONNECTIONS
SO THAT A CHANGE CAN BE
MADE IF NECESSARY TO SUIT
DUCT CONNECTIONS

FOUR MOUNTING HOLES

CEILING DIFFUSER

NECK

SINGLE MOTOR MIXING BOX

COLD (OR HOT) AIR FLOW

ALL FLEXIBLE DUCTS
SHALL BE OF THE PRE-
INSULATED TYPE. INSULATION
SHALL BE AT LEAST 1" THICK
& 3/4 LB. DENSITY GLASS
FIBER BLANKET COMPLETE
WITH EXTERIOR VINYL VAPOR
BARRIER.

PLAN VIEW

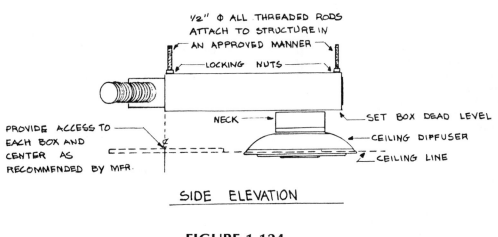

½" Φ ALL THREADED RODS
ATTACH TO STRUCTURE IN
AN APPROVED MANNER

LOCKING NUTS

PROVIDE ACCESS TO
EACH BOX AND
CENTER AS
RECOMMENDED BY MFR.

NECK

SET BOX DEAD LEVEL

CEILING DIFFUSER

CEILING LINE

SIDE ELEVATION

## FIGURE 1-124

## DUAL DUCT HIGH VELOCITY MIXING BOX

Figures 1-124 through 1-133 are standard typical duct work details which
should be incorporated as is where they apply on a drawing or project
requiring any of them. Sizing, equipment notes, and other special items
may be added as required.

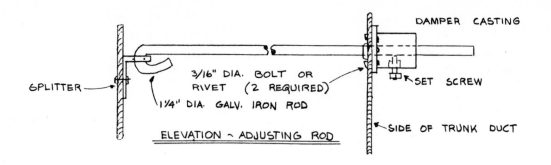

DAMPER CASTING

3/16" DIA. BOLT OR RIVET (2 REQUIRED)

1¼" DIA. GALV. IRON ROD

SPLITTER

SET SCREW

SIDE OF TRUNK DUCT

ELEVATION ~ ADJUSTING ROD

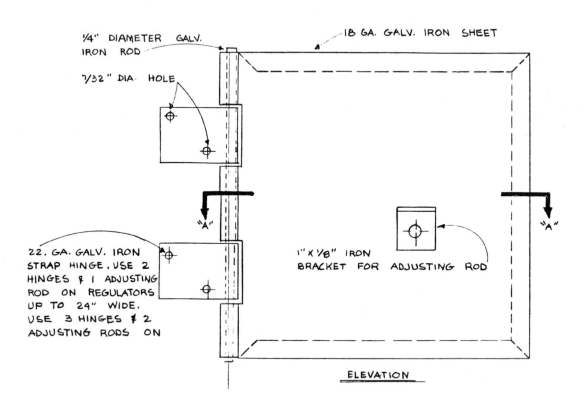

¼" DIAMETER GALV. IRON ROD

7/32" DIA. HOLE

18 GA. GALV. IRON SHEET

22. GA. GALV. IRON STRAP HINGE. USE 2 HINGES & 1 ADJUSTING ROD ON REGULATORS UP TO 24" WIDE. USE 3 HINGES & 2 ADJUSTING RODS ON

"A"

1" X ⅛" IRON BRACKET FOR ADJUSTING ROD

"A"

ELEVATION

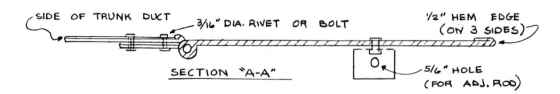

SIDE OF TRUNK DUCT

3/16" DIA. RIVET OR BOLT

½" HEM EDGE (ON 3 SIDES)

SECTION "A-A"

5/6" HOLE (FOR ADJ. ROD)

**FIGURE 1-125**

SPLITTER   TYPE   REGULATOR

130

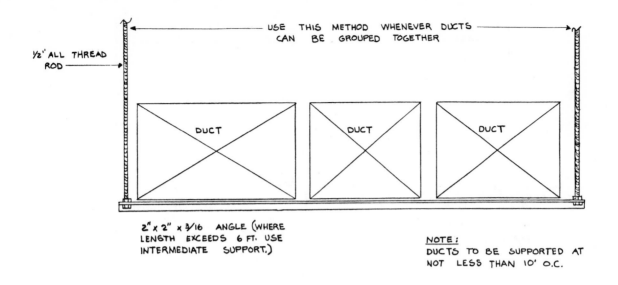

**FIGURE 1-126**

MULTIPLE DUCTS ON TRAPEZE HANGERS

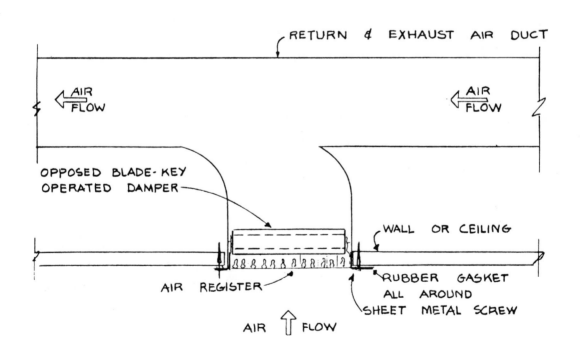

**FIGURE 1-127**

RETURN OR EXHAUST REGISTER INSTALLATION

131

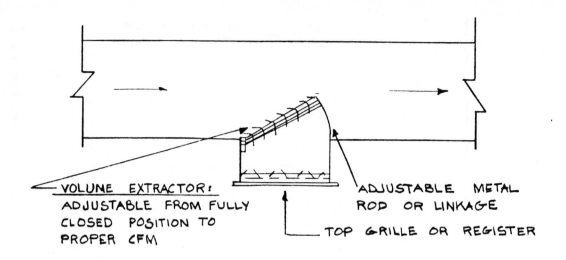

VOLUME EXTRACTOR:
ADJUSTABLE FROM FULLY
CLOSED POSITION TO
PROPER CFM

ADJUSTABLE METAL
ROD OR LINKAGE

TOP GRILLE OR REGISTER

PLAN VIEW

**FIGURE 1-128**

SUPPLY    GRILLE    OR    REGISTER    TAKE-OFF

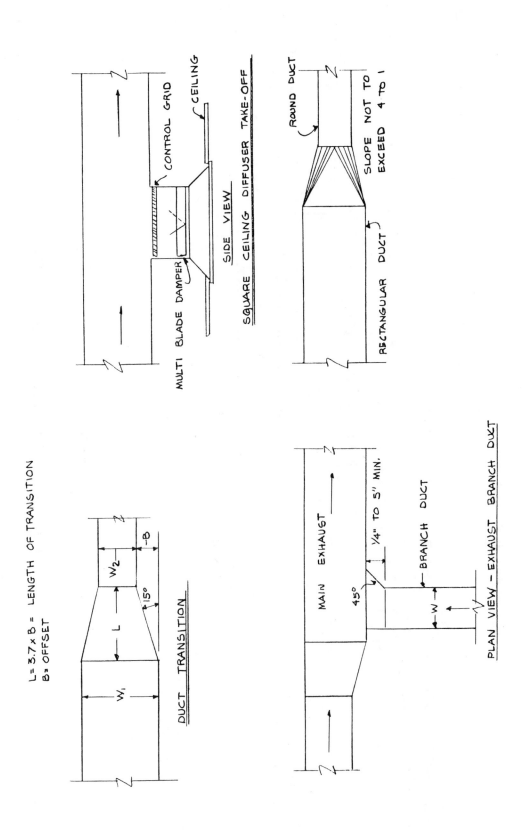

CONTROL GRID

CEILING

MULTI BLADE DAMPER

SIDE VIEW

SQUARE CEILING DIFFUSER TAKE-OFF

ROUND DUCT

SLOPE NOT TO EXCEED 4 TO 1

RECTANGULAR DUCT

L = 3.7 × B = LENGTH OF TRANSITION

B = OFFSET

$W_2$

B

15°

L

$W_1$

DUCT TRANSITION

MAIN EXHAUST

45°

¼" TO 5" MIN.

BRANCH DUCT

W

PLAN VIEW - EXHAUST BRANCH DUCT

**FIGURE 1-129**

DUCT TAKE-OFF & TRANSITION

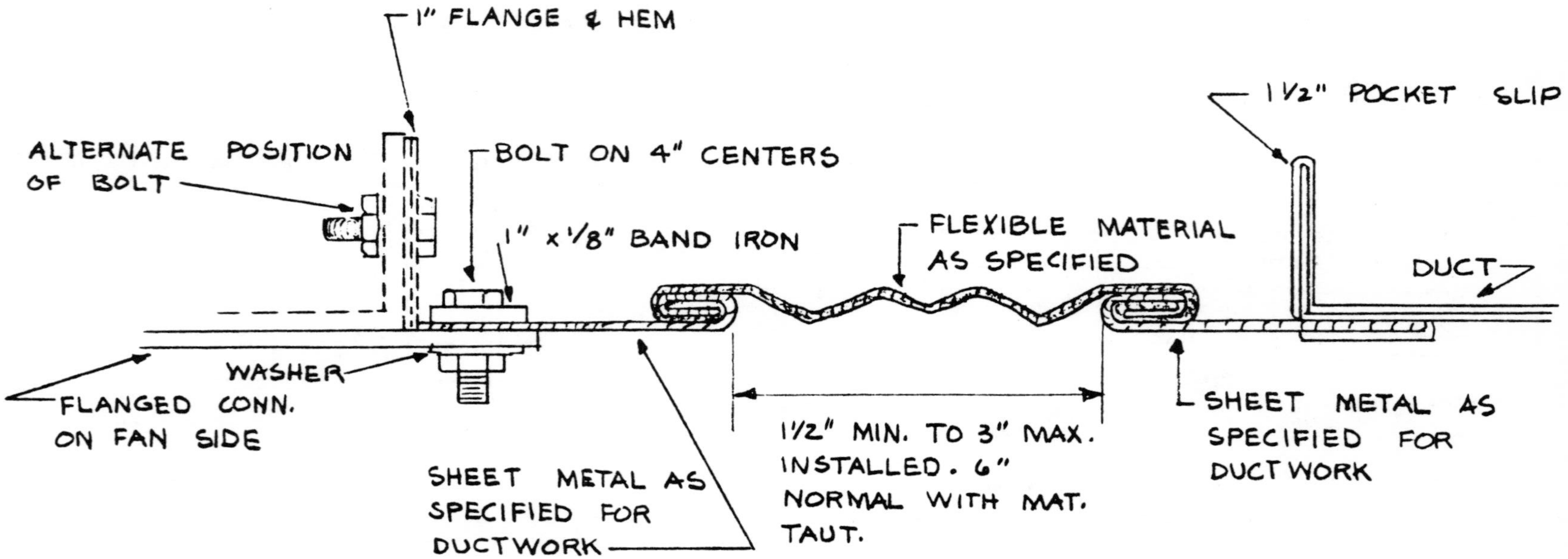
1" FLANGE & HEM
ALTERNATE POSITION OF BOLT
BOLT ON 4" CENTERS
1" x 1/8" BAND IRON
FLEXIBLE MATERIAL AS SPECIFIED
1 1/2" POCKET SLIP
DUCT
WASHER
FLANGED CONN. ON FAN SIDE
SHEET METAL AS SPECIFIED FOR DUCTWORK
1 1/2" MIN. TO 3" MAX. INSTALLED. 6" NORMAL WITH MAT. TAUT.
SHEET METAL AS SPECIFIED FOR DUCTWORK

**FIGURE 1-130**

RECTANGULAR FLEXIBLE CONNECTION

RECTANGULAR DUCT TAKE-OFF SEE FIGURE A,B,C,D FOR ORDER OF PREFERENCE. PROVIDE A SPLITTER DAMPER FOR EACH TAKE OFF.

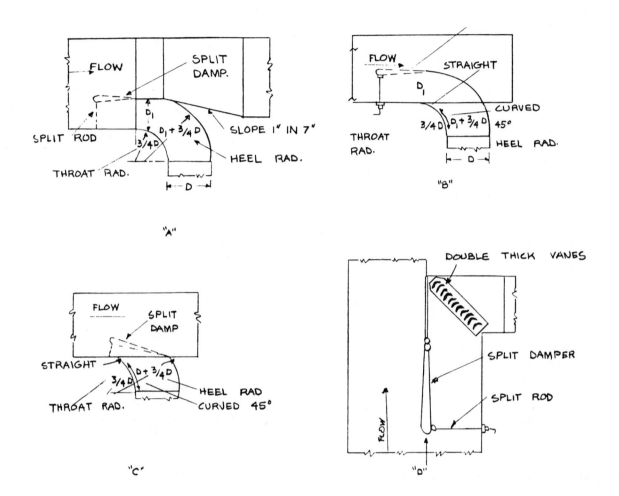

FIGURE 1-131

RECTANGULAR DUCT TAKE-OFF DETAILS

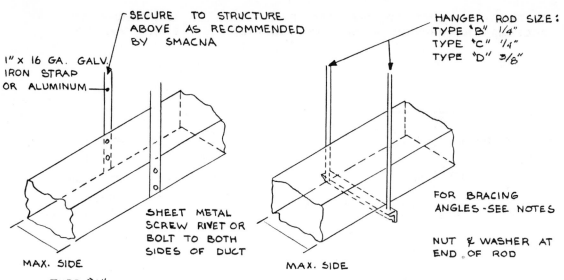

SECURE TO STRUCTURE
ABOVE AS RECOMMENDED
BY SMACNA

1" x 16 GA. GALV.
IRON STRAP
OR ALUMINUM

SHEET METAL
SCREW RIVET OR
BOLT TO BOTH
SIDES OF DUCT

MAX. SIDE

TYPE "A"
8 FT. MAX. HANGER SPACING
ALSO PROVIDE 3 HANGERS AT
EACH TAKE-OFF OR BRANCH

HANGER ROD SIZE:
TYPE "B" 1/4"
TYPE "C" 1/4"
TYPE "D" 3/8"

FOR BRACING
ANGLES-SEE NOTES

NUT & WASHER AT
END OF ROD

MAX. SIDE

TYPE "B" "C" & "D"
8 FT. MAX. HANGER SPACING

| DUCT SCHEDULE | |
|---|---|
| DUCT DIMENSIONS INCHES | TYPE HANGER |
| UP THRU 12 | A |
| 13 18 | A |
| 19 30 | A/B |
| 31 42 | B |
| 43 54 | B |
| 55 60 | B |
| 61 84 | C |
| 85 96 | C |
| OVER 96 | D |

NOTES:

1. FOR SEVERAL DUCTS ON ONE HANGER TYPE "B"-"C"
OR "D" MAY BE USED, SIZE OF HANGER WILL BE
SELECTED ON THE SUM OF DUCT WIDTHS EQUAL
TO MAX. WIDTH OF DUCT SCHEDULE.

2. SCHEDULE FOR ANGLES FOR BRACING: TYPE "B" 1 1/2"
X 1 1/2" X 1/8" ANGLE. MAX. SPACING 8'-0" CENTERS;
TYPE "C" 1 1/2" X 1 1/2" X 3/16" ANGLE MAX SPACING
8'-0" CENTERS; TYPE "D" 2" X 2" X 1/4 MAX SPACING
4'-0" CENTERS.

**FIGURE 1-132**

DUCT HANGERS

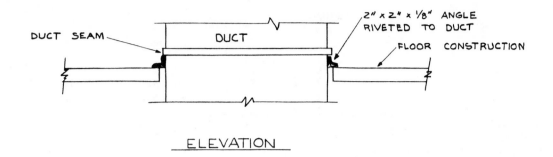

DUCT SEAM

2" x 2" x ⅛" ANGLE RIVETED TO DUCT

FLOOR CONSTRUCTION

DUCT

ELEVATION

NOTE:
ALL DUCTWORK RISERS WHICH ARE RUN EXPOSED, SUCH AS THRU ATTIC, FLOORS AND FAN ROOM FLOORS, SHALL BE PROVIDED WITH A 4" HIGH CONC. CURB AROUND OPENING FOR DUCTS.

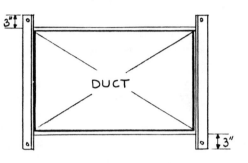

3"

DUCT

3"

PLAN

**FIGURE 1-133**

TYPICAL   SUPPORT   FOR   VERTICAL   DUCTS

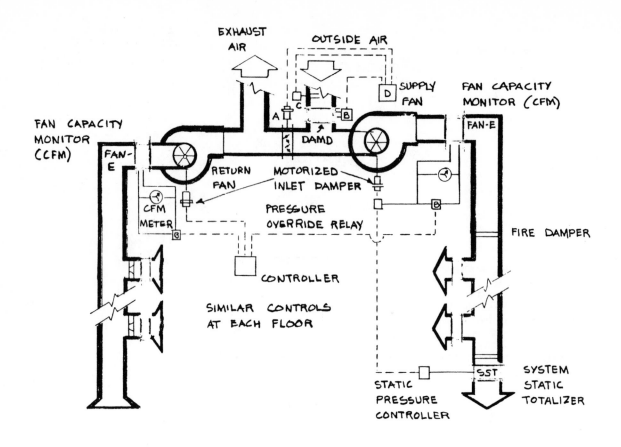

EXHAUST AIR

OUTSIDE AIR

SUPPLY FAN

FAN CAPACITY MONITOR (CFM)

FAN CAPACITY MONITOR (CFM)

FAN-E

FAN-E

A

C

B

D

DAMD

RETURN FAN

MOTORIZED INLET DAMPER

PRESSURE OVERRIDE RELAY

CFM METER

FIRE DAMPER

CONTROLLER

SIMILAR CONTROLS AT EACH FLOOR

STATIC PRESSURE CONTROLLER

SST

SYSTEM STATIC TOTALIZER

A. MOTORIZED VOLUME CONTROL DAMPER
B. VELOCITY PRESSURE TRANSMITTER
C. OUTSIDE AIR THERMOSTAT
D. CONTROLLER

## FIGURE 1-134

## VARIABLE AIR VOLUME SYSTEM CONTROLS

Figures 1-134 and 1-135 are illustrative drawings primarily used to indicate control positioning for constant and variable air volume systems. They are not details in the normal sense, but could become part of a schematic system layout.

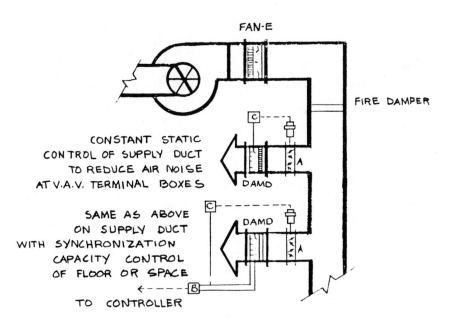

FAN-E

FIRE DAMPER

C

CONSTANT STATIC
CONTROL OF SUPPLY DUCT
TO REDUCE AIR NOISE
AT V.A.V. TERMINAL BOXES

A

DAMD

C

SAME AS ABOVE
ON SUPPLY DUCT
WITH SYNCHRONIZATION
CAPACITY CONTROL
OF FLOOR OR SPACE

DAMD

A

B

TO CONTROLLER

STATIC PRESSURE CONTROL OF BRANCH DUCT FOR
AIR NOISE REDUCTION

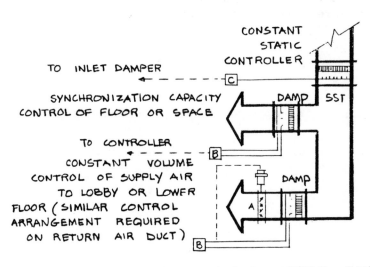

CONSTANT
STATIC
CONTROLLER

TO INLET DAMPER

C

DAMP   SST

SYNCHRONIZATION CAPACITY
CONTROL OF FLOOR OR SPACE

TO CONTROLLER

B

CONSTANT VOLUME
CONTROL OF SUPPLY AIR
TO LOBBY OR LOWER
FLOOR (SIMILAR CONTROL
ARRANGEMENT REQUIRED
ON RETURN AIR DUCT)

DAMP

A

B

A.   MOTORIZED CONTROL DAMPER
B.   VELOCITY PRESSURE TRANSMITTER
C.   STATIC CONTROLLER

CONSTANT VOLUME CONTROL FOR LOBBY & LOWER FLOORS

**FIGURE 1-135**

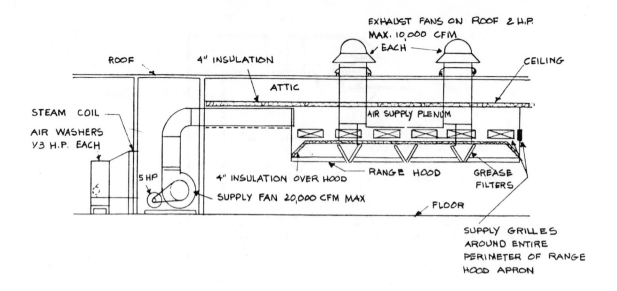

**FIGURE 1-136**

SUPPLY EXHAUST ARRANGEMENT FOR COMMERCIAL
KITCHEN HOODS

Figure 1-136 is an arrangement of supply and exhaust for commercial
kitchen hoods worked out for this particular type of installation and
commonly used in the version presented. As can be seen, CFM arrangements and horsepower sizes were obviously taken from a given job.

140

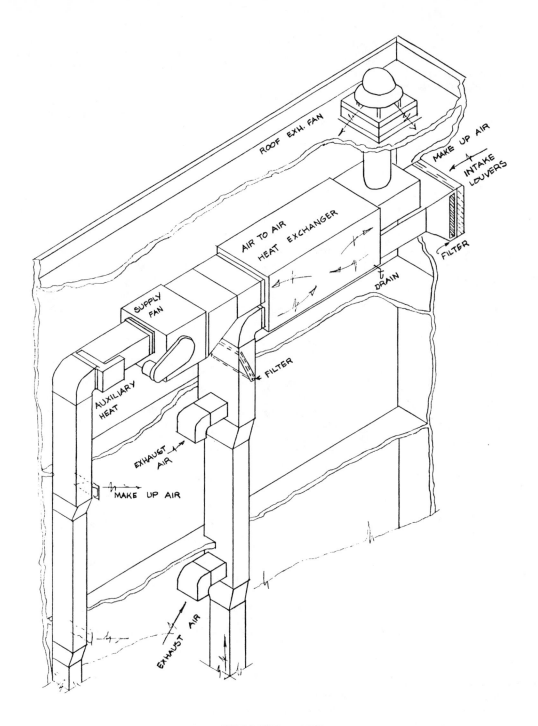

**FIGURE 1-137**

AIR TO AIR TOILET EXHAUST SYSTEM

Figure 1-137 is an arrangement taken from an actual installation showing
an air to air heat exchanger in a toilet exhaust system and is a common
energy conservation method. The design and arrangement is self-evident
in terms of counterflow.

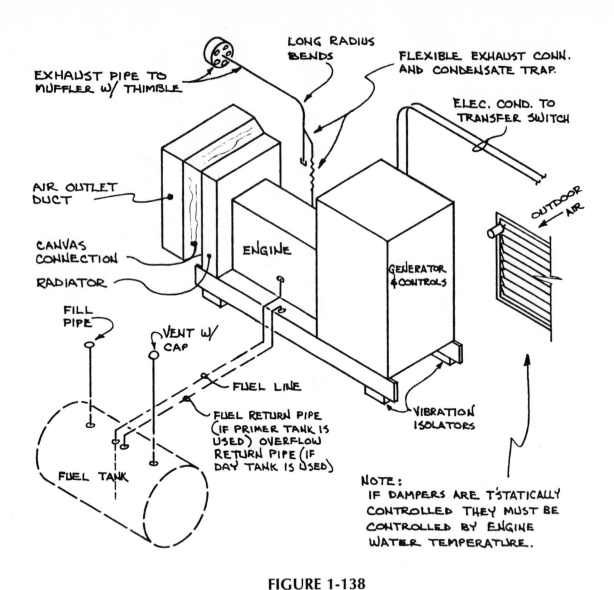

EXHAUST PIPE TO MUFFLER W/ THIMBLE

LONG RADIUS BENDS

FLEXIBLE EXHAUST CONN. AND CONDENSATE TRAP.

ELEC. COND. TO TRANSFER SWITCH

AIR OUTLET DUCT

OUTDOOR AIR

CANVAS CONNECTION

ENGINE

GENERATOR & CONTROLS

RADIATOR

FILL PIPE

VENT W/ CAP

FUEL LINE

FUEL RETURN PIPE (IF PRIMER TANK IS USED) OVERFLOW RETURN PIPE (IF DAY TANK IS USED)

VIBRATION ISOLATORS

FUEL TANK

NOTE:
IF DAMPERS ARE T'STATICALLY CONTROLLED THEY MUST BE CONTROLLED BY ENGINE WATER TEMPERATURE.

## FIGURE 1-138

### TYPICAL GASOLINE ENGINE DRIVEN EMERGENCY GENERATOR W/ RADIATOR COOLING

Figures 1-138 through 1-142 show typical emergency generator installations with various types of fuel and air intake and exhaust. The details are arranged so that each one has slight differences peculiar to the type of cooling and the type of fuel. Care should be taken as to which one is applied to the particular installation.

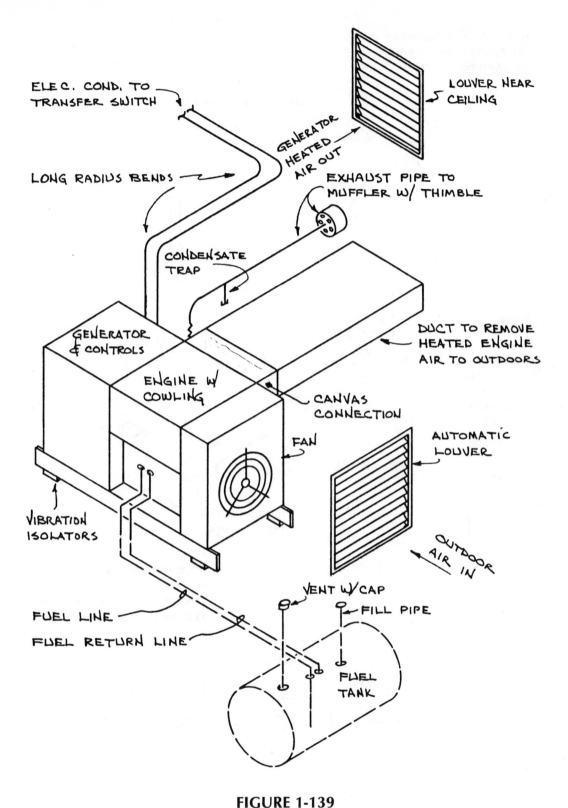

ELEC. COND. TO
TRANSFER SWITCH

LOUVER NEAR
CEILING

GENERATOR
HEATED
AIR OUT

LONG RADIUS BENDS

EXHAUST PIPE TO
MUFFLER W/ THIMBLE

CONDENSATE
TRAP

DUCT TO REMOVE
HEATED ENGINE
AIR TO OUTDOORS

GENERATOR
& CONTROLS

ENGINE W/
COWLING

CANVAS
CONNECTION

FAN

AUTOMATIC
LOUVER

VIBRATION
ISOLATORS

OUTDOOR
AIR IN

VENT W/ CAP

FILL PIPE

FUEL LINE

FUEL RETURN LINE

FUEL
TANK

**FIGURE 1-139**

TYPICAL AIR COOLED (PRESSURE COOLED) ENGINE
DRIVEN EMERGENCY GENERATOR

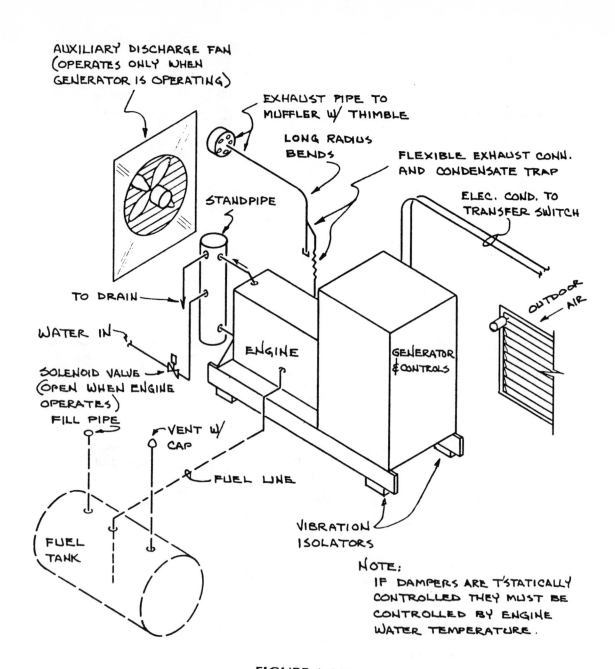

AUXILIARY DISCHARGE FAN
(OPERATES ONLY WHEN
GENERATOR IS OPERATING)

EXHAUST PIPE TO
MUFFLER W/ THIMBLE

LONG RADIUS
BENDS

FLEXIBLE EXHAUST CONN.
AND CONDENSATE TRAP

ELEC. COND. TO
TRANSFER SWITCH

STANDPIPE

TO DRAIN

WATER IN

SOLENOID VALVE
(OPEN WHEN ENGINE
OPERATES)
FILL PIPE

VENT W/
CAP

FUEL LINE

ENGINE

GENERATOR
& CONTROLS

OUTDOOR
AIR

VIBRATION
ISOLATORS

FUEL
TANK

NOTE:
IF DAMPERS ARE T'STATICALLY
CONTROLLED THEY MUST BE
CONTROLLED BY ENGINE
WATER TEMPERATURE.

**FIGURE 1-140**

TYPICAL GASOLINE ENGINE DRIVEN EMERGENCY
GENERATOR W/ CITY WATER COOLING & STANDPIPE

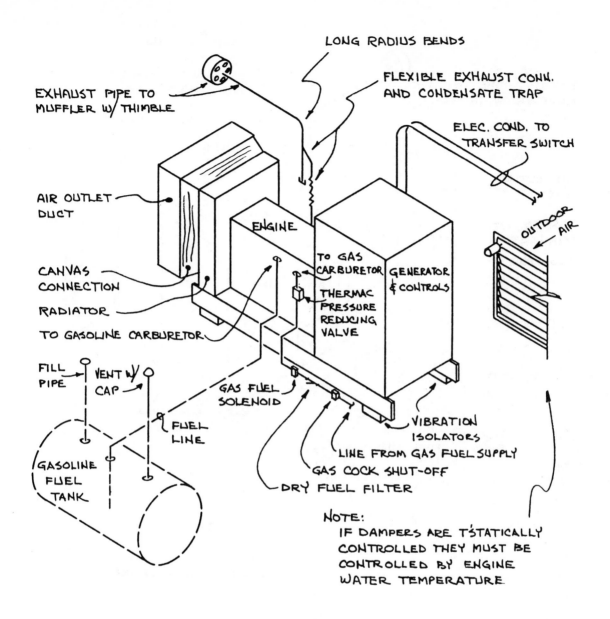

LONG RADIUS BENDS

FLEXIBLE EXHAUST CONN.
AND CONDENSATE TRAP

EXHAUST PIPE TO
MUFFLER W/ THIMBLE

ELEC. COND. TO
TRANSFER SWITCH

AIR OUTLET
DUCT

ENGINE

OUTDOOR AIR

TO GAS
CARBURETOR

GENERATOR
& CONTROLS

CANVAS
CONNECTION

THERMAC
PRESSURE
REDUCING
VALVE

RADIATOR

TO GASOLINE CARBURETOR

FILL
PIPE

VENT W/
CAP

GAS FUEL
SOLENOID

FUEL
LINE

VIBRATION
ISOLATORS

GASOLINE
FUEL
TANK

LINE FROM GAS FUEL SUPPLY

GAS COCK SHUT-OFF

DRY FUEL FILTER

NOTE:
IF DAMPERS ARE T'STATICALLY
CONTROLLED THEY MUST BE
CONTROLLED BY ENGINE
WATER TEMPERATURE

**FIGURE 1-141**

TYPICAL GAS & GASOLINE ENGINE DRIVEN
EMERGENCY GENERATOR W/ RADIATOR COOLING

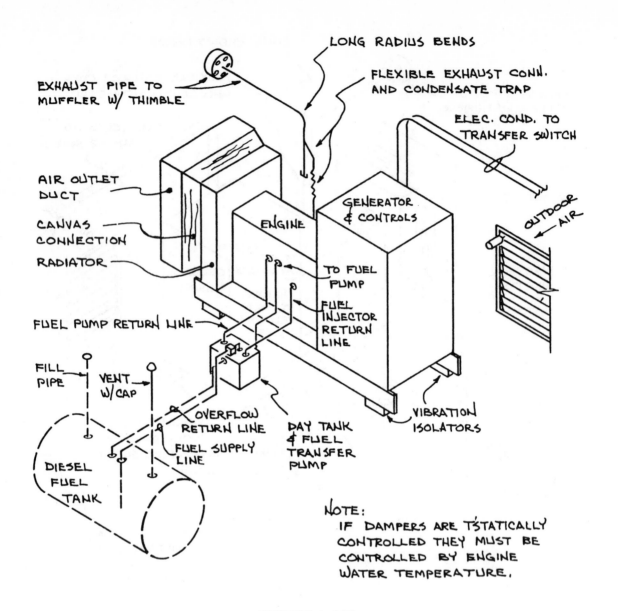

LONG RADIUS BENDS

FLEXIBLE EXHAUST CONN. AND CONDENSATE TRAP

EXHAUST PIPE TO MUFFLER W/ THIMBLE

ELEC. COND. TO TRANSFER SWITCH

AIR OUTLET DUCT

GENERATOR & CONTROLS

ENGINE

OUTDOOR AIR

CANVAS CONNECTION

RADIATOR

TO FUEL PUMP

FUEL INJECTOR RETURN LINE

FUEL PUMP RETURN LINE

FILL PIPE

VENT W/ CAP

OVERFLOW RETURN LINE

FUEL SUPPLY LINE

DAY TANK & FUEL TRANSFER PUMP

VIBRATION ISOLATORS

DIESEL FUEL TANK

NOTE:

IF DAMPERS ARE T'STATICALLY CONTROLLED THEY MUST BE CONTROLLED BY ENGINE WATER TEMPERATURE.

**FIGURE 1-142**

TYPICAL DIESEL ENGINE DRIVEN EMERGENCY GENERATOR W/ RADIATOR COOLING

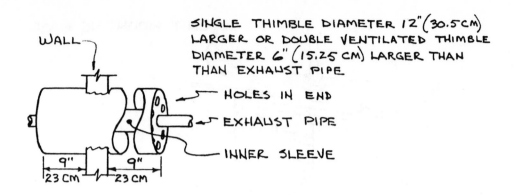

WALL

SINGLE THIMBLE DIAMETER 12" (30.5CM) LARGER OR DOUBLE VENTILATED THIMBLE DIAMETER 6" (15.25 CM) LARGER THAN THAN EXHAUST PIPE

— HOLES IN END

— EXHAUST PIPE

— INNER SLEEVE

9" 23 CM    9" 23 CM

EXHAUST PIPE WITH THIMBLE THRU WALL

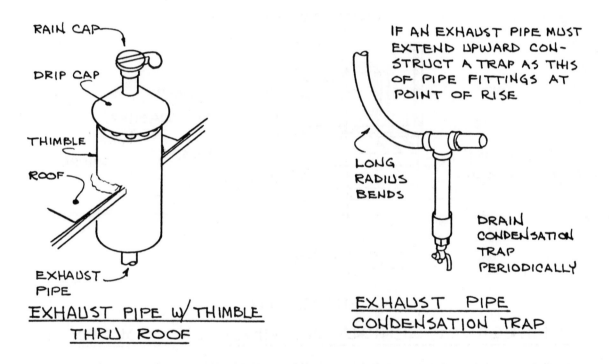

RAIN CAP

DRIP CAP

THIMBLE

ROOF

EXHAUST PIPE

EXHAUST PIPE W/ THIMBLE THRU ROOF

IF AN EXHAUST PIPE MUST EXTEND UPWARD CONSTRUCT A TRAP AS THIS OF PIPE FITTINGS AT POINT OF RISE

LONG RADIUS BENDS

DRAIN CONDENSATION TRAP PERIODICALLY

EXHAUST PIPE CONDENSATION TRAP

## FIGURE 1-143

## MISCELLANEOUS EXHAUST PIPE DETAILS

Figure 1-143 shows details of the exhaust pipe going through the wall or the roof with and without a condensation pipe trap. These details should be carefully checked against local codes before using.

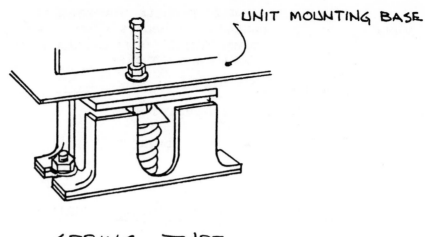

SPRING TYPE

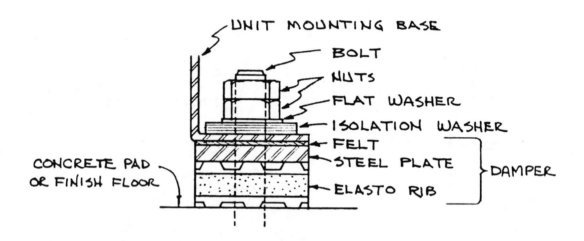

PAD TYPE

**FIGURE 1-144**

EMERGENCY GENERATOR VIBRATION ISOLATORS

Figure 1-144 shows a typical isolation vibrator on an emergency generator. All emergency generators require some form of isolation and these are generally the standard arrangements of isolation and are self-explanatory. While emergency generators are commonly run with diesel fuel and in some cases with natural gas, there are also installations of bottled or low pressure gas.

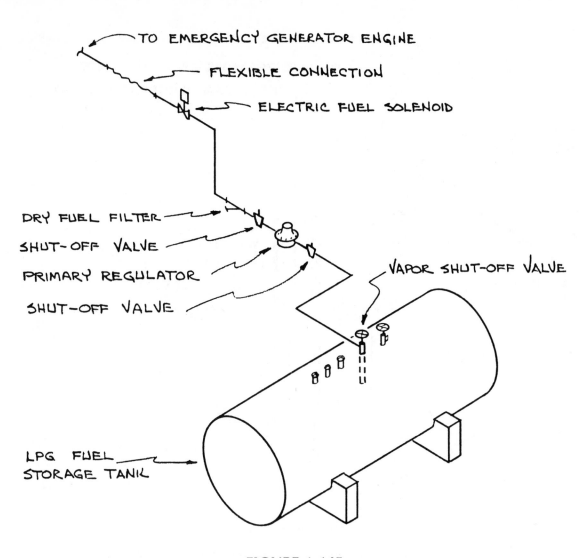

TO EMERGENCY GENERATOR ENGINE

FLEXIBLE CONNECTION

ELECTRIC FUEL SOLENOID

DRY FUEL FILTER

SHUT-OFF VALVE

PRIMARY REGULATOR

SHUT-OFF VALVE

VAPOR SHUT-OFF VALVE

LPG FUEL STORAGE TANK

**FIGURE 1-145**

LPG VAPOR WITHDRAWAL SYSTEM
FOR EMERGENCY GENERATOR

Figures 1-145 and 1-146 are details of the major components of importance in the low pressure gas system.

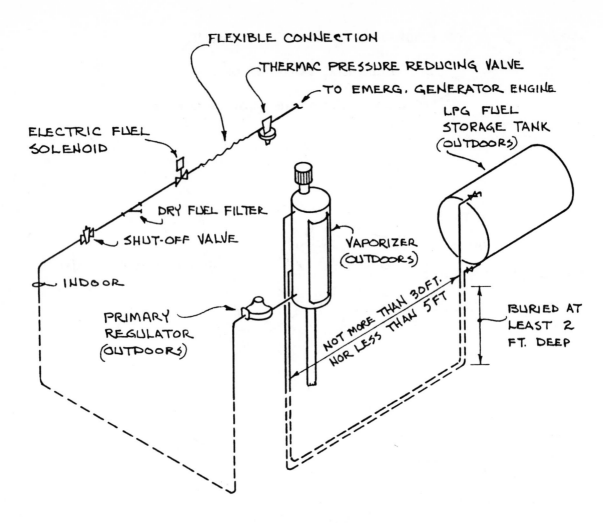

FLEXIBLE CONNECTION

THERMAC PRESSURE REDUCING VALVE

TO EMERG. GENERATOR ENGINE

LPG FUEL STORAGE TANK (OUTDOORS)

ELECTRIC FUEL SOLENOID

DRY FUEL FILTER

SHUT-OFF VALVE

INDOOR

VAPORIZER (OUTDOORS)

PRIMARY REGULATOR (OUTDOORS)

NOT MORE THAN 30FT. NOR LESS THAN 5FT

BURIED AT LEAST 2 FT. DEEP

**FIGURE 1-146**

LPG SYSTEM WITH GAS BURNER VAPORIZER
FOR EMERGENCY GENERATOR

150

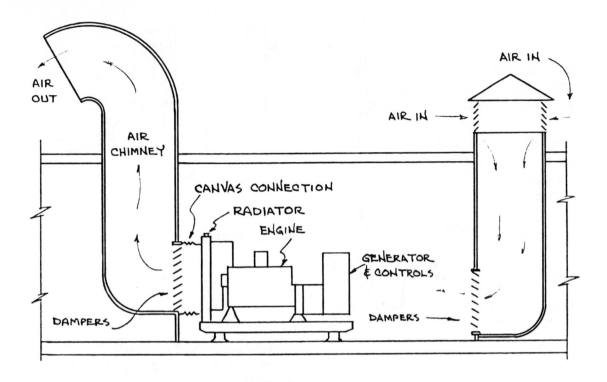

**FIGURE 1-147**

TYPICAL DUCT INSTALLATION WITH ROOFTOP
AIR INLET & OUTLET FOR EMERGENCY GENERATOR

Figure 1-147 shows the generally accepted method of supplying and
exhausting air to the room for the emergency generator. Dampers are an
important part of this installation and other features of the duct work with
regards to wall penetrations should be carefully checked against the local
fire codes.

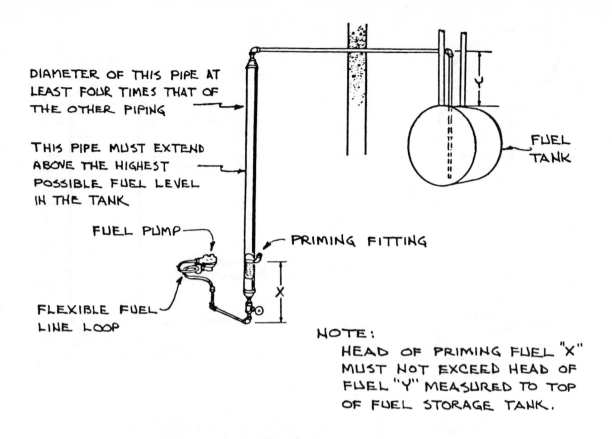

DIAMETER OF THIS PIPE AT LEAST FOUR TIMES THAT OF THE OTHER PIPING

THIS PIPE MUST EXTEND ABOVE THE HIGHEST POSSIBLE FUEL LEVEL IN THE TANK

FUEL PUMP

PRIMING FITTING

FLEXIBLE FUEL LINE LOOP

FUEL TANK

NOTE:
HEAD OF PRIMING FUEL "X" MUST NOT EXCEED HEAD OF FUEL "Y" MEASURED TO TOP OF FUEL STORAGE TANK.

**FIGURE 1-149**

GASOLINE ANTI-SIPHON FUEL SYSTEM FOR EMERGENCY GENERATOR

Figures 1-149 and 1-150 show two possible ways of installing an overhead fuel tank to an emergency generator day tank. One way is common to a certain type of manufacturer's requirements.

152

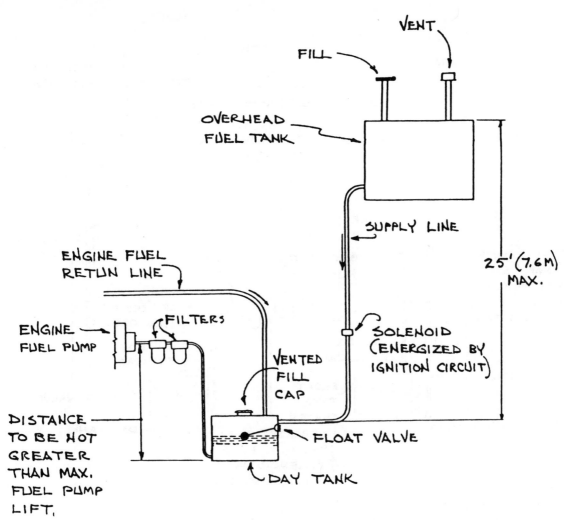

VENT

FILL

OVERHEAD
FUEL TANK

SUPPLY LINE

ENGINE FUEL
RETURN LINE

25' (7.6M)
MAX.

ENGINE
FUEL PUMP

FILTERS

SOLENOID
(ENERGIZED BY
IGNITION CIRCUIT)

VENTED
FILL
CAP

DISTANCE
TO BE NOT
GREATER
THAN MAX.
FUEL PUMP
LIFT.

FLOAT VALVE

DAY TANK

NOTE:
DAY TANK AND LINES
MUST BE BELOW INJECTION
RETURN OUTLET.

FIGURE 1-148

TYPICAL DIESEL FUEL SYSTEM WITH OVERHEAD
FUEL TANK FOR EMERGENCY GENERATOR

Figure 1-148 is an illustration primarily for the designer of an important
part of the anti-siphon system on a gasoline fired emergency generator.
The detail is self-explanatory.

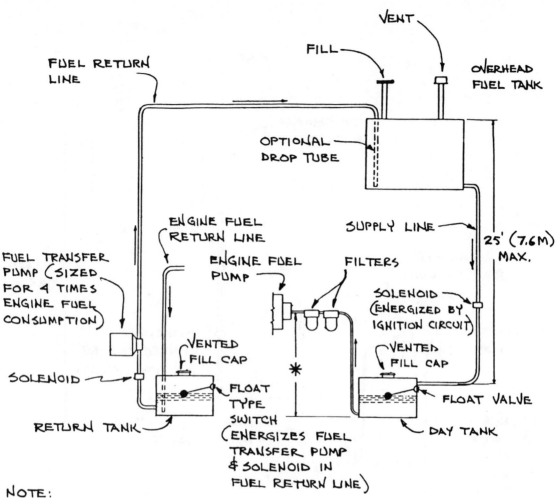

FUEL RETURN LINE

VENT

FILL

OVERHEAD FUEL TANK

OPTIONAL DROP TUBE

ENGINE FUEL RETURN LINE

SUPPLY LINE

25' (7.6M) MAX.

FUEL TRANSFER PUMP (SIZED FOR 4 TIMES ENGINE FUEL CONSUMPTION)

ENGINE FUEL PUMP

FILTERS

SOLENOID (ENERGIZED BY IGNITION CIRCUIT)

SOLENOID

VENTED FILL CAP

VENTED FILL CAP

RETURN TANK

FLOAT TYPE SWITCH (ENERGIZES FUEL TRANSFER PUMP & SOLENOID IN FUEL RETURN LINE)

FLOAT VALVE

DAY TANK

NOTE:
RETURN LINE LIFT MUST NOT EXCEED PUMPING CAPACITY OF TRANSFER PUMP SIZED FOR 4 TIMES ENGINE FUEL CONSUMPTION.

* DISTANCE TO BE NOT GREATER THAN MAX. FUEL PUMP LIFT.

**FIGURE 1-150**

TYPICAL DETROIT DIESEL FUEL SYSTEM WITH OVERHEAD TANK FOR EMERGENCY GENERATOR

154

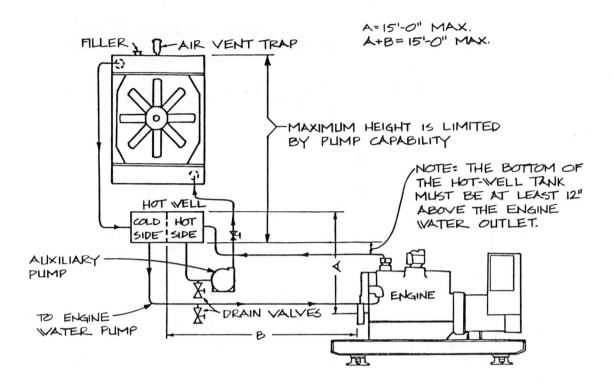

A=15'-0" MAX.
A+B=15'-0" MAX.

FILLER — AIR VENT TRAP

MAXIMUM HEIGHT IS LIMITED BY PUMP CAPABILITY

NOTE: THE BOTTOM OF THE HOT-WELL TANK MUST BE AT LEAST 12" ABOVE THE ENGINE WATER OUTLET.

HOT WELL

COLD SIDE | HOT SIDE

AUXILIARY PUMP

ENGINE

TO ENGINE WATER PUMP

DRAIN VALVES

A

B

**FIGURE 1-151**

HIGH REMOTE RADIATOR INSTALLATION

NO SCALE

Figures 1-151 and 1-152 illustrate the general schematic arrangement of an emergency generator with a remote radiator, both with and without surge tank. These are guidance drawings and require additional refinement for a particular application.

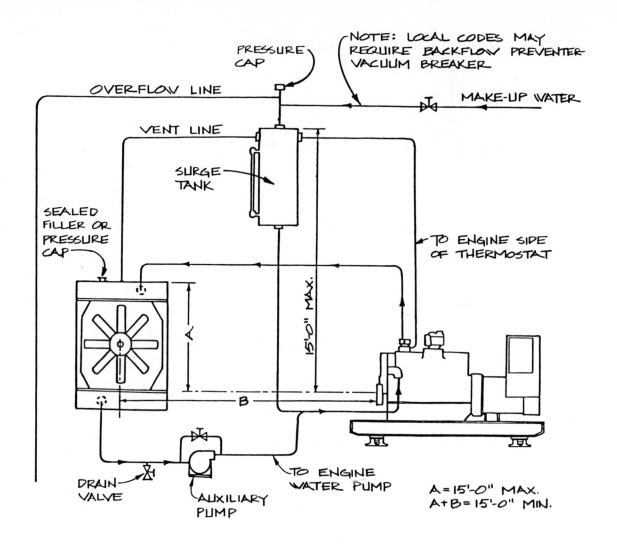

**FIGURE 1-152**

LONG REMOTE RADIATOR INSTALLATION
WITH SURGE TANK

NO SCALE

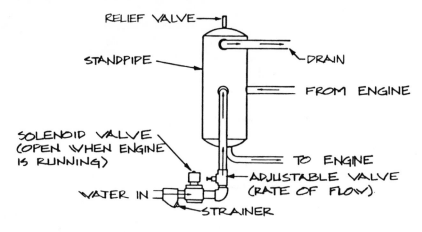

RELIEF VALVE

STANDPIPE

DRAIN

FROM ENGINE

SOLENOID VALVE
(OPEN WHEN ENGINE
IS RUNNING)

TO ENGINE

ADJUSTABLE VALVE
(RATE OF FLOW)

WATER IN

STRAINER

## STANDPIPE COOLING SYSTEM

NO SCALE

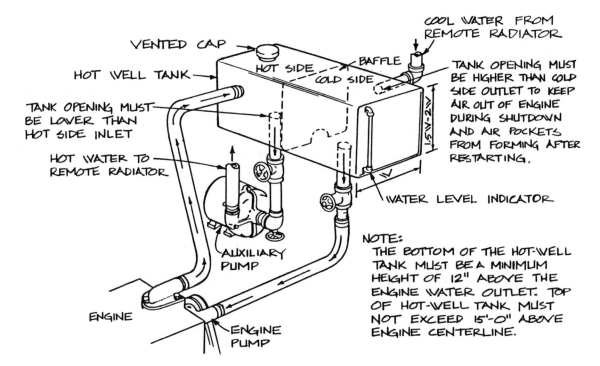

VENTED CAP

COOL WATER FROM
REMOTE RADIATOR

HOT WELL TANK

HOT SIDE

BAFFLE

COLD SIDE

TANK OPENING MUST
BE HIGHER THAN COLD
SIDE OUTLET TO KEEP
AIR OUT OF ENGINE
DURING SHUTDOWN
AND AIR POCKETS
FROM FORMING AFTER
RESTARTING.

TANK OPENING MUST
BE LOWER THAN
HOT SIDE INLET

HOT WATER TO
REMOTE RADIATOR

15'-V-2'-V

WATER LEVEL INDICATOR

AUXILIARY
PUMP

ENGINE

ENGINE
PUMP

NOTE:
THE BOTTOM OF THE HOT-WELL
TANK MUST BE A MINIMUM
HEIGHT OF 12" ABOVE THE
ENGINE WATER OUTLET. TOP
OF HOT-WELL TANK MUST
NOT EXCEED 15'-0" ABOVE
ENGINE CENTERLINE.

**FIGURE 1-153**

## TYPICAL HOT WELL TANK

NO SCALE

Figures 1-153, 1-154, and 1-155 indicate certain specific details which
require sizing in order to be incorporated in the drawing. Instead of
sizing, certain notes that are pertinent to these particular details are
shown.

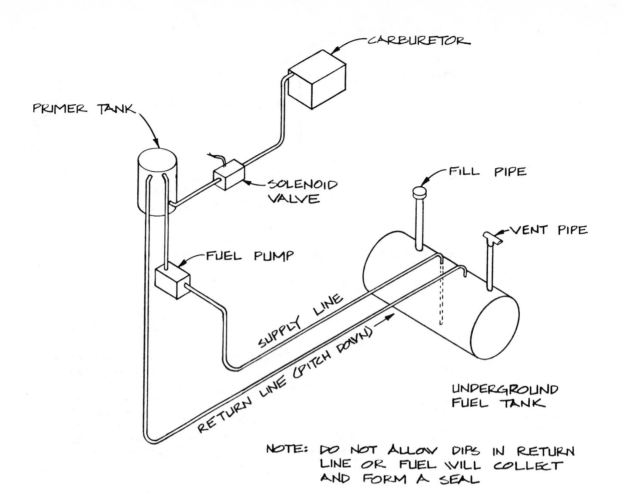

CARBURETOR

PRIMER TANK

SOLENOID VALVE

FILL PIPE

VENT PIPE

FUEL PUMP

SUPPLY LINE

RETURN LINE (PITCH DOWN)

UNDERGROUND FUEL TANK

NOTE: DO NOT ALLOW DIPS IN RETURN LINE OR FUEL WILL COLLECT AND FORM A SEAL

**FIGURE 1-154**

GASOLINE PRIMER TANK SYSTEM

NO SCALE

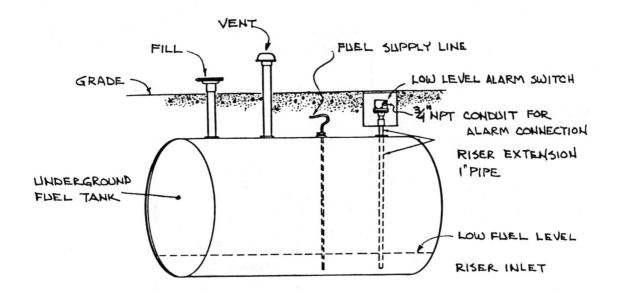

FILL

VENT

GRADE

FUEL SUPPLY LINE

LOW LEVEL ALARM SWITCH

3/4" NPT CONDUIT FOR ALARM CONNECTION

RISER EXTENSION 1" PIPE

UNDERGROUND FUEL TANK

LOW FUEL LEVEL

RISER INLET

NOTE:
ALLOW FOR FUEL LEVEL DROP WITHIN 1" (25.4 MM) OF RISER INLET BEFORE SWITCH ACTUATION. ADD 1" FOR EACH 10" (OR 8.3MM/MTR) OF VERTICAL RISER PIPE EXTENSION.

**FIGURE 1-155**

LOW LEVEL ALARM SWITCH
FOR FUEL TANK

159

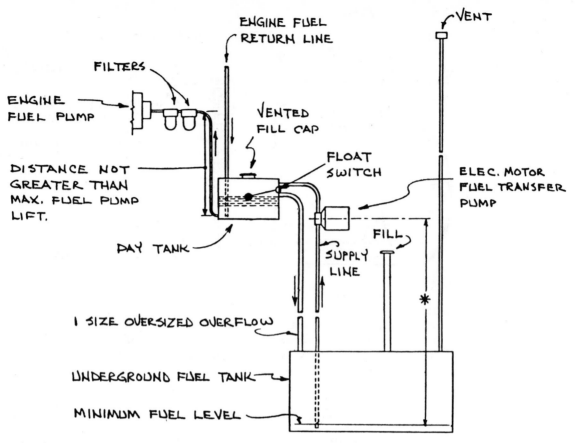

ENGINE FUEL RETURN LINE

FILTERS

ENGINE FUEL PUMP

VENTED FILL CAP

VENT

FLOAT SWITCH

DISTANCE NOT GREATER THAN MAX. FUEL PUMP LIFT.

ELEC. MOTOR FUEL TRANSFER PUMP

DAY TANK

FILL

SUPPLY LINE

*

1 SIZE OVERSIZED OVERFLOW

UNDERGROUND FUEL TANK

MINIMUM FUEL LEVEL

NOTE:
   DAY TANK AND LINES
   MUST BE BELOW
   INJECTOR RETURN OUTLET.

*  DISTANCE MUST NOT BE
   GREATER THAN FUEL
   TRANSFER PUMP LIFT
   CAPACITY.

**FIGURE 1-156**

TYPICAL DIESEL FUEL SYSTEM WITH
SUPPLY TANK BELOW GENERATOR SET

Figures 1-156 and 1-157 are the reverse of the previous details with tanks above the generator set. In these cases, the tanks are below the generator set and in both drawings, the cautionary asterisk indicates the distance, which must be carefully calculated.

160

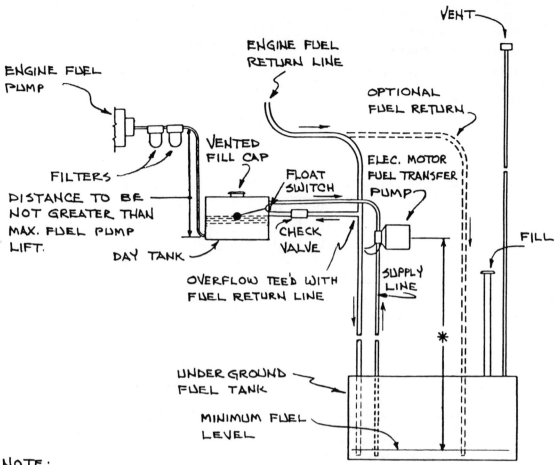

ENGINE FUEL
PUMP

ENGINE FUEL
RETURN LINE

OPTIONAL
FUEL RETURN

VENT

FILTERS

VENTED
FILL CAP

FLOAT
SWITCH

ELEC. MOTOR
FUEL TRANSFER
PUMP

DISTANCE TO BE
NOT GREATER THAN
MAX. FUEL PUMP
LIFT.

CHECK
VALVE

FILL

DAY TANK

SUPPLY
LINE

OVERFLOW TEE'D WITH
FUEL RETURN LINE

UNDER GROUND
FUEL TANK

MINIMUM FUEL
LEVEL

NOTE:
   DAY TANK AND LINES
   MUST BE BELOW
   INJECTOR RETURN OUTLET.

✳ DISTANCE MUST NOT BE
   GREATER THAN FUEL
   TRANSFER PUMP LIFT
   CAPACITY.

**FIGURE 1-157**

TYPICAL DETROIT DIESEL FUEL SYSTEM WITH
SUPPLY TANK BELOW GENERATOR SET

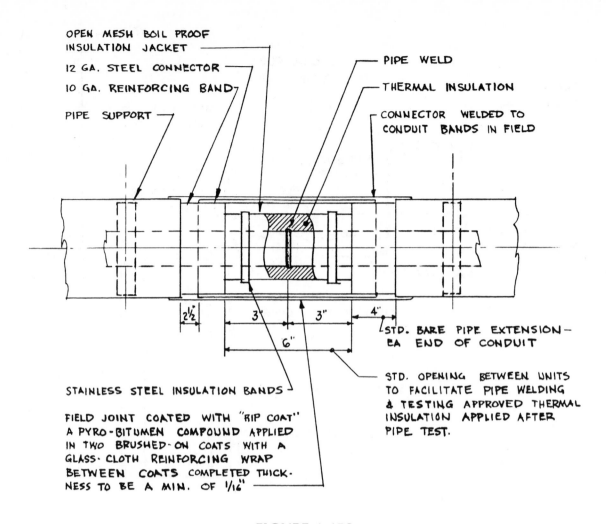

OPEN MESH BOIL PROOF
INSULATION JACKET

12 GA. STEEL CONNECTOR

10 GA. REINFORCING BAND

PIPE SUPPORT

PIPE WELD

THERMAL INSULATION

CONNECTOR WELDED TO
CONDUIT BANDS IN FIELD

2½"    3"    3"    4"

6"

STD. BARE PIPE EXTENSION —
EA END OF CONDUIT

STAINLESS STEEL INSULATION BANDS

FIELD JOINT COATED WITH "RIP COAT"
A PYRO-BITUMEN COMPOUND APPLIED
IN TWO BRUSHED-ON COATS WITH A
GLASS-CLOTH REINFORCING WRAP
BETWEEN COATS COMPLETED THICK-
NESS TO BE A MIN. OF 1/16"

STD. OPENING BETWEEN UNITS
TO FACILITATE PIPE WELDING
& TESTING APPROVED THERMAL
INSULATION APPLIED AFTER
PIPE TEST.

FIGURE 1-158

DETAIL OF FIELD JOINT FOR UNDERGROUND
CONDUIT

NO SCALE

Figures 1-158 through 1-163 indicate standard details for underground
prefabricated conduits. These details include the conduit itself, the
method of going through a wall, the anchoring method, the expansion
joint, and the general sectional view of a steam supply and return in the
conduit. The steam supply and return could be any number of pipes of
any material or system that is planned for this type of conduit.

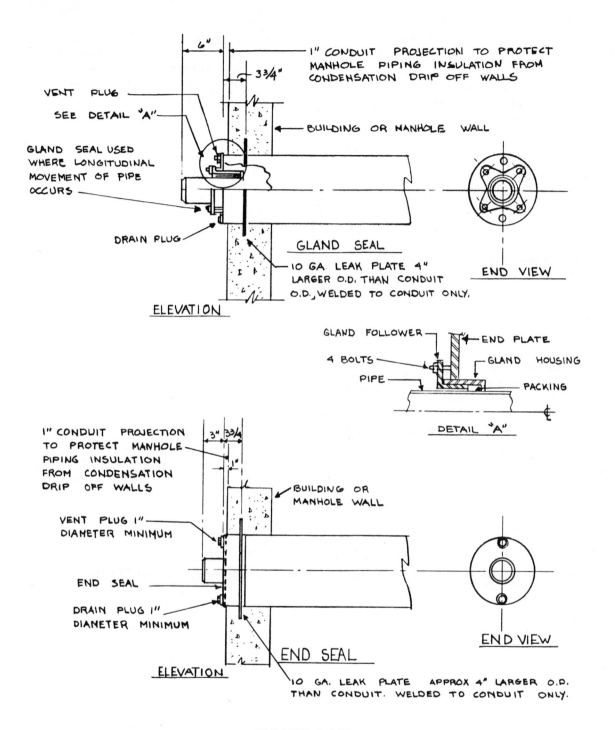

VENT PLUG
SEE DETAIL "A"

GLAND SEAL USED WHERE LONGITUDINAL MOVEMENT OF PIPE OCCURS

DRAIN PLUG

6"

3¾"

1" CONDUIT PROJECTION TO PROTECT MANHOLE PIPING INSULATION FROM CONDENSATION DRIP OFF WALLS

BUILDING OR MANHOLE WALL

GLAND SEAL

10 GA. LEAK PLATE 4" LARGER O.D. THAN CONDUIT O.D., WELDED TO CONDUIT ONLY.

ELEVATION

END VIEW

GLAND FOLLOWER

4 BOLTS

PIPE

END PLATE

GLAND HOUSING

PACKING

DETAIL "A"

1" CONDUIT PROJECTION TO PROTECT MANHOLE PIPING INSULATION FROM CONDENSATION DRIP OFF WALLS

VENT PLUG 1" DIAMETER MINIMUM

END SEAL

DRAIN PLUG 1" DIAMETER MINIMUM

3" 3¾"

1"

BUILDING OR MANHOLE WALL

END SEAL

ELEVATION

END VIEW

10 GA. LEAK PLATE APPROX 4" LARGER O.D. THAN CONDUIT. WELDED TO CONDUIT ONLY.

**FIGURE 1-159**

PREFABRICATED UNDERGROUND CONDUIT
DETAILS OF BUILDING WALL ENTRY

NO SCALE

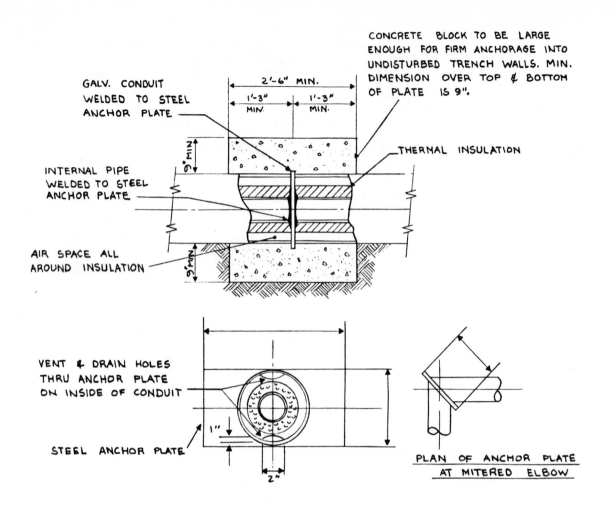

GALV. CONDUIT WELDED TO STEEL ANCHOR PLATE

INTERNAL PIPE WELDED TO STEEL ANCHOR PLATE

AIR SPACE ALL AROUND INSULATION

CONCRETE BLOCK TO BE LARGE ENOUGH FOR FIRM ANCHORAGE INTO UNDISTURBED TRENCH WALLS. MIN. DIMENSION OVER TOP & BOTTOM OF PLATE IS 9".

THERMAL INSULATION

2'-6" MIN.
1'-3" MIN.
1'-3" MIN.
9" MIN
9" MIN

VENT & DRAIN HOLES THRU ANCHOR PLATE ON INSIDE OF CONDUIT

STEEL ANCHOR PLATE

1"
2"

PLAN OF ANCHOR PLATE AT MITERED ELBOW

**FIGURE 1-160**

PREFABRICATED UNDERGROUND CONDUIT
DETAILS OF STANDARD ANCHOR CONSTRUCTION

NO SCALE

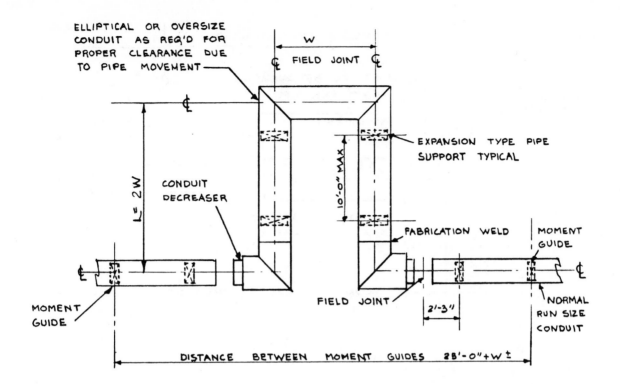

**FIGURE 1-161**

DETAIL OF PPEFABRICATED UNDERGROUND CONDUIT
FABRICATED EXPANSION LOOP

NO SCALE

165

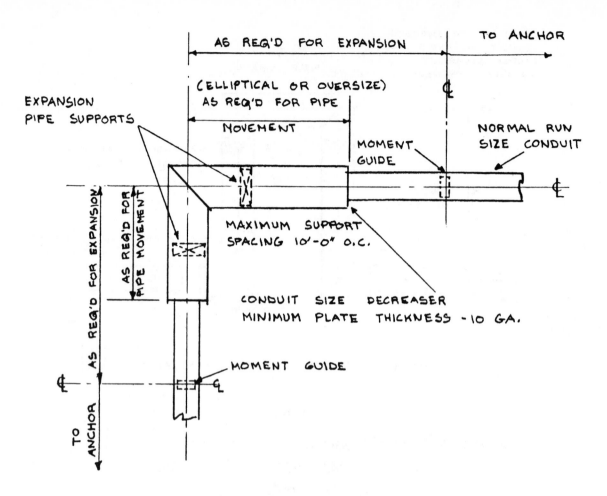

AS REQ'D FOR EXPANSION

TO ANCHOR

(ELLIPTICAL OR OVERSIZE)
AS REQ'D FOR PIPE
MOVEMENT

EXPANSION
PIPE SUPPORTS

MOMENT
GUIDE

NORMAL RUN
SIZE CONDUIT

AS REQ'D FOR EXPANSION

AS REQ'D FOR PIPE MOVEMENT

MAXIMUM SUPPORT
SPACING 10'-0" O.C.

CONDUIT SIZE DECREASER
MINIMUM PLATE THICKNESS - 10 GA.

MOMENT GUIDE

TO ANCHOR

**FIGURE 1-162**

DETAIL OF PREFABRICATED UNDERGROUND CONDUIT
EXPANSION ELBOW

NO SCALE

166

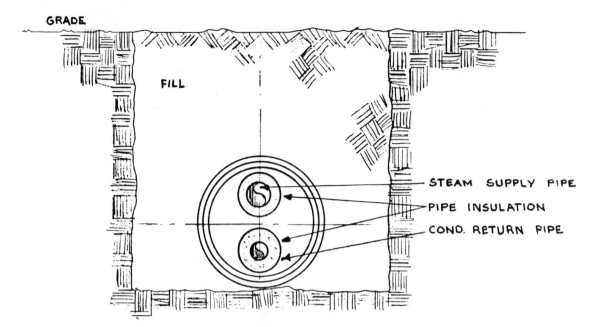

CROSS SECTION THRU PIPE TRENCH

STEAM SUPPLY PIPE
PIPE INSULATION
COND. RETURN PIPE

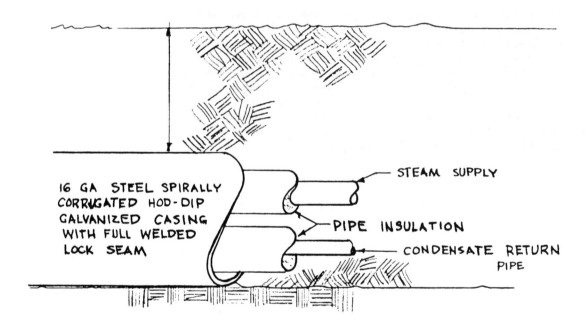

16 GA STEEL SPIRALLY CORRUGATED HOD-DIP GALVANIZED CASING WITH FULL WELDED LOCK SEAM

STEAM SUPPLY
PIPE INSULATION
CONDENSATE RETURN PIPE

LONGITUDINAL SECTION THRU PIPE TRENCH

**FIGURE 1-163**

DETAILS OF PREFABRICATED UNDERGROUND CONDUIT

NO SCALE

167

| AIR CONDITIONING DESIGN DATA | | | | | | | |
|---|---|---|---|---|---|---|---|
| DESIGN AREA | SUMMER | | | | WINTER | | |
| | OUTSIDE | | INSIDE | | OUTSIDE | INSIDE | |
| | DB | WB | DB | WB | DB | DB | %HUMIDITY |
| ALL AREAS | 92 | 74 | 73 | 61 | -11 | 73 | 50 |
| | | | | | | | |

| MAJOR AIR CONDITIONING EQUIPMENT ITEMS | | | | | |
|---|---|---|---|---|---|
| EQUIPMENT | LOCATION | PERFORMANCE MINIMUM CAPACITY | MOTOR | | REMARKS |
| | | | NOM. HP | PHASE VOLT. | |
| | | | | | |
| | | | | | |

# FIGURE 1-164

MECHANICAL   SCHEDULES

Figures 1-164 through 1-183 are heating and ventilating details that are dedicated to mechanical schedules. Mechanical schedules as such are anything the designer wants them to be. All of this material could be included in the specification and none of it shown on the plans. Both ways should be tried and an evaluation made, based on the circumstances and the requirements of the client as well as individual opinion of what can be better understood if it were shown on the plans. They are not necessarily what are wanted. They have information that could be easily omitted. There are other bits of information to each schedule that could be added. These are simply a version of a reasonable coverage of each of the subjects presented.

| | HEATING-VENTILATING & AIR CONDITIONING UNIT FANS (BUILT UP SYSTEM) | | | | | | | | | |
|---|---|---|---|---|---|---|---|---|---|---|
| UNIT NO. | LOCATION | FAN CFM | O.A. CFM | FAN ARRANGEMENT | FAN MOTOR | | TOTAL S.P. OF SYSTEM | TYPE WHEEL | MIN. WHEEL DIA. |
| | | | | | NOM. H.P. | PHASE-VOLT. | | | |
| | | | | | | | | | |
| | | | | | | | | | |
| | | | | | | | | | |

| | PREFILTERS | | | | | |
|---|---|---|---|---|---|---|
| CFM | SYSTEM | MAX. FACE VELOCITY | TYPE | MIN. EFFICIENCY NBS COTTRELL TEST | MAX. CLEAN FILTER SP | CALCULATED SP FOR SYSTEM |
| | | | | | | |
| | | | | | | |

| | FINAL FILTERS | | | | | | |
|---|---|---|---|---|---|---|---|
| CFM | SYSTEM | CARTRIDGES | | | MIN. EFFICIENCY NBS. ATMOSPHERIC DUST SPOT TEST | MAX. CLEAN FILTER SP | CALCULATED SP FOR SYSTEM |
| | | NUMBER | SIZE | ARRANGEMENT | | | |
| | | | | | | | |
| | | | | | | | |

## FIGURE 1-165

MECHANICAL SCHEDULES

| | CIRCULATING PUMPS | | | | | | | | |
|---|---|---|---|---|---|---|---|---|---|
| LOCATION | SYSTEM | CFM | HEAD FT. | FLUID TEMP °F. | % EFF. | DOUBLE OR SINGLE SUCTION | NOM. HP | MOTOR PHASE-VOLT. | RPM |
| | | | | | | | | | |
| | | | | | | | | | |

| | CHILLED WATER COOLING COILS | | | | | | | | | | | | |
|---|---|---|---|---|---|---|---|---|---|---|---|---|---|
| COIL ROWS | SYSTEM | CFM | MAX. FACE VEL. FPM. | MAX. S.P. | ENT.AIR °F. | | LVG. AIR °F. | | CIRC. WATER | | | | BTUH |
| | | | | | DB. | WB. | DB. | WB. | TEMP. IN °F | TEMP. OUT °F | GPM | MAX. P.D. FT.WATER | |
| | | | | | | | | | | | | | |
| | | | | | | | | | | | | | |

| | STEAM HUMIDIFIER | | | | | | | | |
|---|---|---|---|---|---|---|---|---|---|
| HUMID. NO. | SYSTEM | CFM | WINTER | | | | STEAM PSIG | CONTROL VALVE LBS/HR | TRAP LBS/HR |
| | | | ENT. AIR °F. | | LVG. AIR °F. | | | | |
| | | | DB. | WB. | DB. | WB. | | | |
| | | | | | | | | | |
| | | | | | | | | | |

## FIGURE 1-166

MECHANICAL SCHEDULES

| STEAM HUMIDIFIER | | | | | | | |
|---|---|---|---|---|---|---|---|
| HUMID. NO. | SYSTEM | CFM | WINTER | | STEAM PSIG | CONTROL VALVE LBS/HR | TRAP LBS/HR |
| | | | ENT. AIR. °F. | LVG. AIR. °F. | | | |
| | | | | | | | |
| | | | | | | | |

| PREHEAT COIL | | | | | | | | | | | |
|---|---|---|---|---|---|---|---|---|---|---|---|
| COIL NO. | SYSTEM | CFM | MAX. FACE VEL.FPM | MAX. S.P. | TEMP. AIR °F | | MIN. BTUH | STEAM PSIG | | CONTROL VALVE LBS/HR | TRAP LBS/HR |
| | | | | | ENT. | LVG. | | ENTERING CONTROL VALVE | ENTERING COIL | | |
| | | | | | | | | | | | |
| | | | | | | | | | | | |

| PROPELLOR OR AXIAL FLOW EXHAUST FANS | | | | | | | | | | | |
|---|---|---|---|---|---|---|---|---|---|---|---|
| FAN NO. | OUT.VEL. F.P.M. | CFM | RPM | S.P. | TIP SPEED | HP | VOLTS | PHASE | DAMPER TYPE | WHEEL DIAM. | REMARKS |
| | | | | | | | | | | | |
| | | | | | | | | | | | |

## FIGURE 1-167

MECHANICAL SCHEDULES

| MOTOR AND MOTOR CONTROL SCHEDULE | | | | | | | | | | | | | | | | |
|---|---|---|---|---|---|---|---|---|---|---|---|---|---|---|---|---|
| MOTOR | LOCATION | HP | VOLTS | PHASE | AMPS | | STARTER | | | | IND. LTS. | CONTROL VOLTAGE | MANUAL CONTROL | | AUTOMATIC CONT. | |
| | | | | | FL. | LR. | TYPE | NEMA SIZE | CTR. BKR. | MTG. | | | DEVICE | LOCATION | | |
| | | | | | | | | | | | | | | | | |
| | | | | | | | | | | | | | | | | |

MATCH LINE

| | INTERLOCKING | | | | |
|---|---|---|---|---|---|
| SAFETY DEVICE | NO.CONTACTS | | ITEM | REMARKS |
| | N.O. | N.C. | | |
| | | | | |
| | | | | |

MATCH LINE

| SCHEDULE OF LOUVER SIZES | | | | | |
|---|---|---|---|---|---|
| LOUVER No. | WIDTH | HEIGHT | MATERIAL | SCREEN | REMARKS |
| | | | | | |
| | | | | | |

## FIGURE 1-168

MECHANICAL SCHEDULES

| | | | HEAT RECLAIMER COILS | | | | |
|---|---|---|---|---|---|---|---|
| SYSTEM | CFM | S.P. | MAX. FACE VEL.FPM | AIR. TEMP. °F. | | MEDIUM | NOTE |
| | | | | ENT. | LVG. | | |
| | | | | | | | |
| | | | | | | | |

| | | | | | REHEAT COIL | | | | | | |
|---|---|---|---|---|---|---|---|---|---|---|---|
| COIL. NO. | AREA SERVED | CFM | MAX. FACE VEL.FPM | MAX. S.P. | TEMP. AIR °F. | | MIN. BTUH | STEAM PSIG | | CONTROL VALVE LBS/HR | TRAP LBS/HR |
| | | | | | ENT. | LVG. | | ENTERING CONTROL VALVE | ENTERING COIL | | |
| | | | | | | | | | | | |
| | | | | | | | | | | | |

## FIGURE 1-169

MECHANICAL SCHEDULES

| | | | | CAPACITY SCHEDULE FOR THE VERTICAL AIR HANDLING UNIT WITH | | | | | | | | | | MATCH LINE |
|---|---|---|---|---|---|---|---|---|---|---|---|---|---|---|
| UNIT NO. | CFM | EXT SP | MAX. O.V. | FAN MOTOR | | | | MIN.COIL FACE AREA | WATER FLOW THRU COIL GPM | MAX.WATER PD THRU COIL FT | COIL CAPACITY - HEATING | | | |
| | | | | HP | VOLTS | PHASE | CYCLE | | | | ENT. AIR °F. | LV.AIR °F. | BTU/HR | |
| | | | | | | | | | | | | | | |

| | | FLAT FILTER BOX AND COMBINATION COIL | | | | | | |
|---|---|---|---|---|---|---|---|---|
| MATCH LINE | COIL CAPACITY - COOLING | | | HEATING WATER | | COOLING WATER | | |
| | ENT.AIR °FDB °FWB | LV.AIR °FDB °FWB | BTU/HR | ENTERS | LEAVES | ENTERS | LEAVES | |
| | | | | | | | | |
| | | | | | | | | |

## FIGURE 1-170

MECHANICAL SCHEDULES

| SCHEDULE OF CAPACITIES OF THE VERTICAL HOT WATER UNIT HEATER | | | | | | | | | | | | |
|---|---|---|---|---|---|---|---|---|---|---|---|---|
| | | | FAN MOTOR | | | | | | WATER | | | |
| UNIT HEATER NO. | CFM | R.P.M. | H.P. | VOLTS | PHASE | CYCLE | BTU/HR. CAPACITY | G.P.M. | ENTERS °F | LEAVES°F | MAX.PRESS.DROP | |
| | | | | | | | | | | | | |
| | | | | | | | | | | | | |

MATCH LINE

| | AIR | | MOUNTING HEIGHT | THROW IN FEET |
|---|---|---|---|---|
| | ENTER °F. | LEAVES °F. | | |
| | | | | |
| | | | | |

(MATCH LINE)

## FIGURE 1-171

MECHANICAL SCHEDULES

| SCHEDULE OF CAPACITIES OF THE VERTICAL STEAM UNIT HEATER | | | | | | | | | | |
|---|---|---|---|---|---|---|---|---|---|---|
| | | | | | | | | STEAM | | |
| UNIT HEATER NO | C.F.M. | R.P.M. | H.P. | VOLTS | PHASE | CYCLE | BTU/HR.CAPACITY | STEAM PRESS | LBS. OF STEAM/HR. | |
| | | | | | | | | | | |
| | | | | | | | | | | |

MATCH LINE

| | AIR | | MOUNTING HEIGHT | THROW IN FT. |
|---|---|---|---|---|
| | ENTER °F. | LEAVE °F. | | |
| | | | | |
| | | | | |

(MATCH LINE)

## FIGURE 1-172

MECHANICAL SCHEDULES

| SCHEDULE OF CAPACITIES FOR THE AIR COOLED CONDENSER | | | | | | | | | |
|---|---|---|---|---|---|---|---|---|---|
| CONDENSER NUMBER | CFM | STATIC PRESS | NO. OF FANS | NO.FAN MOT. | H.P. | VOLTS | PHASE | GROSS HEAT OR REJECTION | MAX. AMB. TEMP. |
| | | | | | | | | | |
| | | | | | | | | | |

MATCH LINE

| | MINIMUM AMB. TEMP. | CONDENSING TEMP. | TYPE OF LOW AMB.CONTROL | APPROX. WEIGHT |
|---|---|---|---|---|
| | | | | |
| | | | | |

(MATCH LINE)

## FIGURE 1-173

MECHANICAL SCHEDULES

172

| SCHEDULE OF CAPACITIES OF THE WALL HUNG INDUCTION UNIT, COMBINATION COIL WITH WATER CONTROL | | | | | | | | | | | | |
|---|---|---|---|---|---|---|---|---|---|---|---|---|
| | | NOZZ. | COOLING CAPACITY | | | HEATING CAPACITY | | WATER FLOW | | | | |
| SYMBOL | CFM | PRESS | SENS.CAP. | PRI.AIR CAP | TOTAL CAP. | AIR-ON CAP. | GRAVITY CAP | GPM | MAX.PD | C.W. ENT. | C.W.LVS. | MATCH LINE |
| | | | | | | | | | | | | |
| | | | | | | | | | | | | |
| | | | | | | | | | | | | |

| MATCH LINE | | | | | | |
|---|---|---|---|---|---|---|
| | WATER FLOW | | CONNECTION SIZES | | | |
| | H.W.ENTERS | H.W.LEAVES | PRIM. AIR | WATER SUP. | WATER RET. | DRAIN |
| | | | | | | |
| | | | | | | |

**FIGURE 1-174**

MECHANICAL SCHEDULES

| DUAL COIL FAN COIL UNIT SCHEDULE | | | | | | | | | | | |
|---|---|---|---|---|---|---|---|---|---|---|---|
| | | MOTOR | MAX. | COOLING COIL CAPACITY | | | | | | | |
| SYM. | CFM | HP | WATTS | GPM | MAX.PD | SENSIBLE CAP. | LATENT CAP. | TOTAL CAP. | WATER ENT. | WATER LVS. | MATCH LINE |
| | | | | | | | | | | | |
| | | | | | | | | | | | |

| MATCH LINE | | | | | | | |
|---|---|---|---|---|---|---|---|
| | HEATING COOLING CAPACITY | | | | | RUNOUTS & VALVE SIZE | |
| | GPM | MAX. P.D. | HEATING CAP. | WATER ENT. | WATER LVS. | COOLING | HEATING |
| | | | | | | | |
| | | | | | | | |

**FIGURE 1-175**

MECHANICAL SCHEDULES

| | | | HEATING | | | | | | | MOTOR | | | FAN | | PIPE RUNOUT SZ | | | |
|---|---|---|---|---|---|---|---|---|---|---|---|---|---|---|---|---|---|---|
| TOTAL CFM | RET. AIR CFM | MAX. FRESH AIR C.F.M. | AIR | | HOT WATER | | | | | MOTOR | | | MAX. O.V. | MAX. RPM | S. | R. | DR. | REMARKS |
| | | | ENT. | LVG | ENT. | LVG. | GPM | P.D | BTU/HR | MIN HP | ∅ | MAX. RPM | | | | | | |
| | | | | | | | | | | | | | | | | | | |
| | | | | | | | | | | | | | | | | | | |

SCHEDULE OF CAPACITIES OF THE CLASSROOM TYPE UNIT VENTILATOR, HOT WATER HEATING

| | | | | | FIN TUBE RADIATION SCHEDULE | | | | | | | | |
|---|---|---|---|---|---|---|---|---|---|---|---|---|---|
| SYMBOL | BTU/HR | GPM | WATER ENTERS | WATER LEAVES | HEATING ELEMENT | | | | ENCLOSURE | | RUNOUT SIZES | VALVE SIZES | REMARKS |
| | | | | | TUBE | FINS | FINS LENGTH | ROWS | D | H | | | |
| | | | | | | | | | | | | | |
| | | | | | | | | | | | | | |

| | | | | FIN TUBE RADIATION SCHEDULE | | | | | | | | | |
|---|---|---|---|---|---|---|---|---|---|---|---|---|---|
| SYMBOL | BTU/HR | STEAM PSI | HEATING ELEMENT | | | | ENCLOSURE | | RUNOUT SIZES | | VALVE SIZE | TRAP SIZE | REMARKS |
| | | | TUBE | FINS | FINS LENGTH | ROWS | D | H | STEAM | RET. | | | |
| | | | | | | | | | | | | | |

## FIGURE 1-176

MECHANICAL SCHEDULES

| | | | | FAN - COIL UNIT SCHEDULE | | | | | | | | |
|---|---|---|---|---|---|---|---|---|---|---|---|---|
| | ENTERING WATER TEMPERATURE | | | | | | | TEMPERATURE DROP | | | | |
| LOCATION | QUANTITY | SIZE | TYPE | HEATING CAPACITY | S & R | CFM | O.V. | S.P. | RPM | HP | NOTES | |
| | | | | | | | | | | | | |
| | | | | | | | | | | | | |

| FRESH AIR INTAKE SCHEDULE | | | | |
|---|---|---|---|---|
| UNIT NUMBER | THROAT SIZE | THROAT AREA | MAX. CFM | MAX. THROAT VEL. |
| | | | | |
| | | | | |

## FIGURE 1-177

MECHANICAL SCHEDULES

| SCHEDULE OF CAPACITIES FOR THE STEAM TO WATER CONVERTOR | | | | | | | | |
|---|---|---|---|---|---|---|---|---|
| CONVERTOR NUMBER | WATER SIDE | | | | | STEAM SIDE | | |
| | G.P.M. | WATER ENT. °F. | WATER LVS. °F. | BTU/HR | MAX.PD.FT. | PRESSURE | LBS/Hr | TRAP CAP |
| | | | | | | | | |
| | | | | | | | | |

| SCHEDULE OF CAPACITIES FOR THE WATER TO WATER CONVERTOR | | | | | | | | | |
|---|---|---|---|---|---|---|---|---|---|
| CONVERTOR NUMBER | HIGHER TEMPERATURE WATER SIDE | | | | | LOWER TEMPERATURE WATER SIDE | | | |
| | G.P.M. | WATER ENTERS °F. | WATER LVS. °F. | BTU/HR | MAX.PD.FT. | G.P.M. | WAT.ENT.°F. | WAT.LV.°F | BTU/HR |
| | | | | | | | | | |
| | | | | | | | | | |

## FIGURE 1-178

MECHANICAL SCHEDULES

| SCHEDULE OF CAPACITIES OF THE DUAL DUCT AIR HANDLING UNIT | | | | | | | | | | | | | MATCH LINE |
|---|---|---|---|---|---|---|---|---|---|---|---|---|---|
| UNIT NO. | TOTAL CFM | RETURN CFM | OUTDOOR CFM | MAX. CFM COLD DECK | MAX. CFM HOT DECK | COOLING COIL | | | | | | | |
| | | | | | | AIR ENT | AIR LVS | BTU/HR | GPM | WAT. ENT. | WAT. LVS. | MAX WATER PD | |
| | | | | | | | | | | | | | |
| | | | | | | | | | | | | | |

| MATCH LINE | HEATING COIL | | | | | | | TOT.STATIC PRESSURE INCH. W.G. | FAN MOTOR | | | |
|---|---|---|---|---|---|---|---|---|---|---|---|---|
| | AIR ENT. | AIR LVS. | BTU/HR | GPM | WAT. ENT. | WAT. LVS. | MAX.WAT. P.D. | | HP | VOLTS | PHASE | CYCLE |
| | | | | | | | | | | | | |
| | | | | | | | | | | | | | |

## FIGURE 1-179

MECHANICAL SCHEDULES

| UNIT NO. | CFM | EXT. SP | MAX. OV. | FAN MOTOR | | | | | HEATING COIL CAPACITY | | | | | | | |
|---|---|---|---|---|---|---|---|---|---|---|---|---|---|---|---|---|
| | | | | HP | VOLTS | Ø | CYCLE | MIN.COIL FACE.AREA | AIR ENT.°F. | AIR LV.°F | BTU/HR | GPM | MAX WATER PD. | WAT.ENT. | WAT.LVS. |
| | | | | | | | | | | | | | | | |

*(title row:)* CAPACITY SCHEDULE FOR THE VERTICAL AIR HANDLING UNIT WITH FLAT FILTER BOX & DUAL COILS — MATCH LINE

| COOLING COIL CAPACITY | | | | | | | | |
|---|---|---|---|---|---|---|---|---|
| MIN.COIL FACE AREA | AIR ENT.DB.WB. | AIR LV.DB.WB. | BTU/HR | GPM | MAX WATER PD. | WATER ENTERS | WATER LVS. | MAX AIR VEL. THRU FILTER. |
| | | | | | | | | |

FIGURE 1-180

MECHANICAL   SCHEDULES

| GRILLE AND DIFFUSER SCHEDULE | | | | | | |
|---|---|---|---|---|---|---|
| DESIGNATION | MODEL | SIZE | CFM | TYPE | FINISH | DUCT.SIZE |
| | | | | | | |
| | | | | | | |

| SCHEDULE OF MIXING BOXES | | | | | |
|---|---|---|---|---|---|
| SYMBOL | CFM | MIN. Ps | MAX. Ps | SOUND RATING N℃ | REMARKS |
| | | | | | |
| | | | | | |

FIGURE 1-181

MECHANICAL   SCHEDULES

| SCHEDULE OF CAPACITIES OF THE ROOF MOUNTED UTILITY VENT SET | | | | | | | | | |
|---|---|---|---|---|---|---|---|---|---|
| FAN NUMBER | CFM | S.P. | MOTOR | | | | MAX. OUTLET VELOCITY | OUTLET VELOCITY | FAN SERVES |
| | | | H.P. | VOLTS | PHASE | CYCLE | | | |
| | | | | | | | | | |

| EXHAUST FAN SCHEDULE | | | | | | | | | | | | |
|---|---|---|---|---|---|---|---|---|---|---|---|---|
| FAN NO. | LOCATION | SERVE | CFM | S.P. | MAX RPM | MIN. H.P. | TIP SPEED | DIR. DRIVE OR V-BELT | TYPE LOUVER | VOLTS | CYCLE | FAN SERVES |
| | | | | | | | | | | | | |
| | | | | | | | | | | | | |

## FIGURE 1-182

MECHANICAL SCHEDULES

| SCHEDULE OF CAPACITIES OF THE SUPPLY AND RETURN LIGHT FIXTURE | | | | | | |
|---|---|---|---|---|---|---|
| PLAN SYMBOL | FIXT. DIMEN. | PROTOTYPE | NO. TUBES & SIZE | SEN. HEAT BTU/HR. TO RM. | CFM SUPPLY AIR | CFM OF RET. AIR |
| | | | | | | |
| | | | | | | |

MATCH LINE

| MATCH LINE | SEN. HEAT BTU/HR. TO RETURN | COMMENTS |
|---|---|---|
| | | |
| | | |

## FIGURE 1-183

MECHANICAL SCHEDULES

# 2. Plumbing Systems

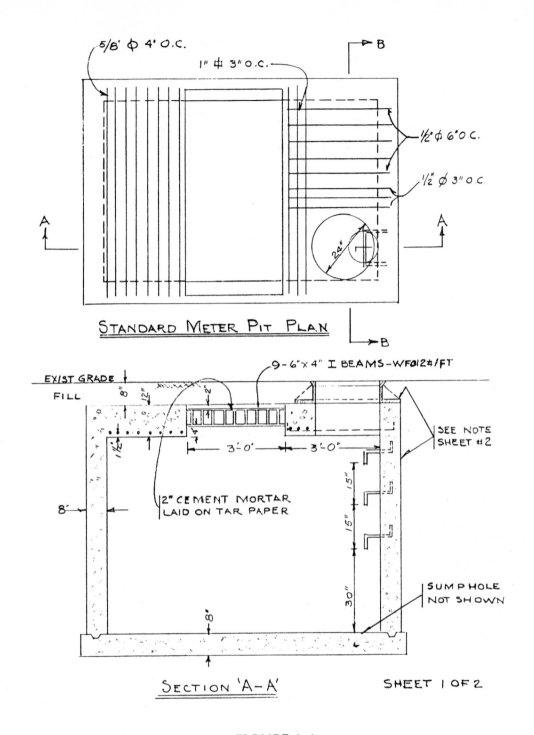

STANDARD METER PIT PLAN

SECTION 'A-A'

SHEET 1 OF 2

**FIGURE 2-1**

Figures 2-1 and 2-2 indicate a standard water meter pit detail. The pit conforms to local standards and generally applies almost anywhere. The actual length, width, and height of the pit depends on the particular application. Note that if the pit is to be constructed under a road or where there is to be any heavy load, the design of the pit is not sufficient to withstand this loading. A special structural design for the pit would be required.

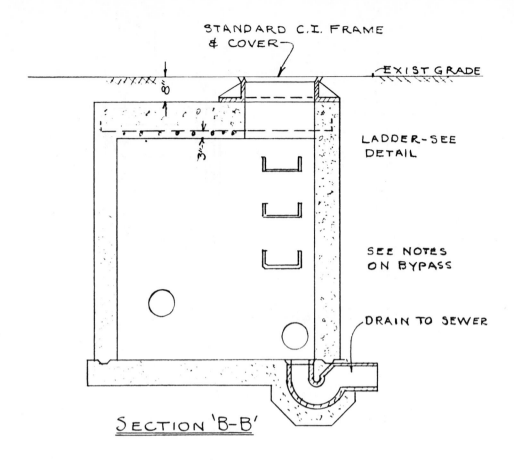

STANDARD C.I. FRAME & COVER

EXIST GRADE

LADDER-SEE DETAIL

SEE NOTES ON BYPASS

DRAIN TO SEWER

SECTION 'B-B'

NOTES
1. PIT NOT TO BE USED UNDER ROADS OR WHERE HEAVY LOADS ARE IMPOSED.
2. WATER PROOF TOP & SIDES OF PIT WITH 2 PLY TAR PAPER & TAR OR CONSTRUCT OF WATER PROOF CEMENT
3. BYPASS MAY BE CONSTRUCTED OUTSIDE & AROUND PIT WITH DRIP EXTENDED THRU WALL. BYPASS MUST ALWAYS BRANCH TO RIGHT LOOKING ALONG DIRECTION OF FLOW

SHEET 2 OF 2

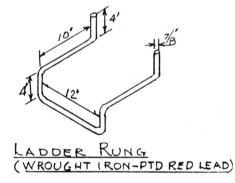

LADDER RUNG
(WROUGHT IRON-PTD RED LEAD)

**FIGURE 2-2**

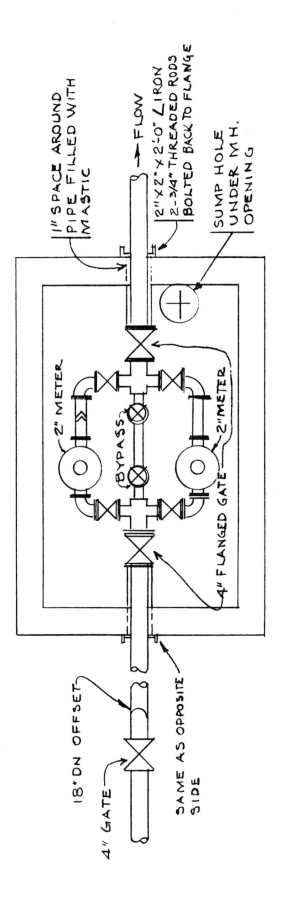

**FIGURE 2-3**

4" Service — Dual Meters

— Not to Scale —

Figures 2-3 and 2-4 are standard 150-lb cement line, cast-iron, water service meter pit installations. The 4" is a typical size and there are two types of metering. Compound metering is generally not well accepted and should be checked out carefully with the local utility company before using the design shown on Figure 2-4. Note that the sleeve stiffener is a requirement.

183

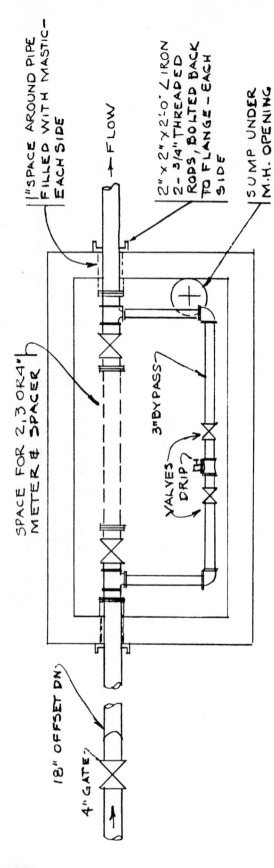

1" SPACE AROUND PIPE FILLED WITH MASTIC— EACH SIDE

→ FLOW

2"×2"×2'-0" ∠ IRON 2-3/4" THREADED RODS, BOLTED BACK TO FLANGE—EACH SIDE

SUMP UNDER M.H. OPENING

SPACE FOR 2, 3 OR 4" METER & SPACER

3" BY PASS

VALVES

DRIP

18" OFFSET DN

4" GATE

**FIGURE 2-4**

4" SERVICE - 2, 3 OR 4" COMPOUND METER

—— NOT TO SCALE ——

184

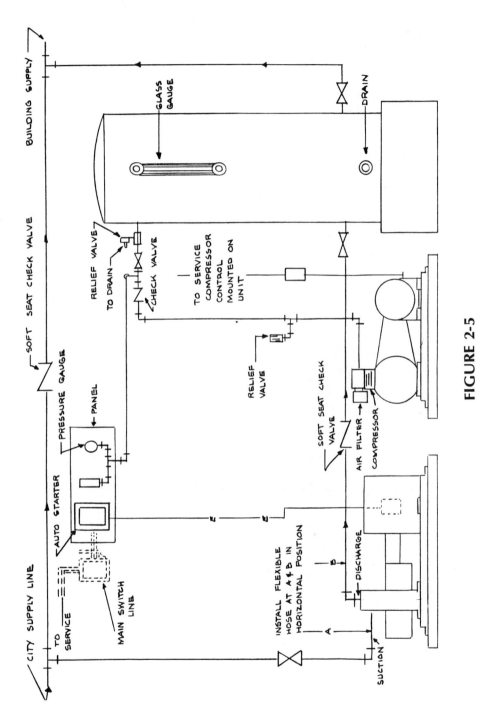

**FIGURE 2-5**

PNEUMATIC WATER SYSTEM

Figure 2-5 is a typical installation where the water supply is insufficient for the pressure required by the system designed. The detail is relatively self-explanatory. The size of the system is a function of load requirements.

185

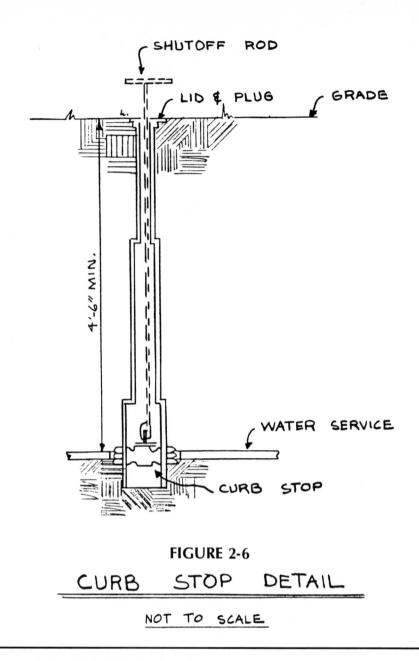

**FIGURE 2-6**

## CURB STOP DETAIL

NOT TO SCALE

Figure 2-6 is a standard curb shutoff. The shutoff must terminate in some sort of curb box at grade. This can be specified or shown in the detail.

186

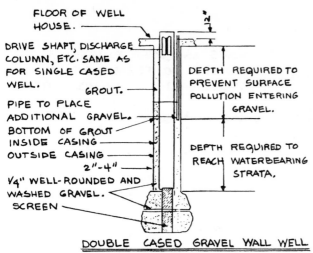

FLOOR OF WELL HOUSE.

DRIVE SHAFT, DISCHARGE COLUMN, ETC. SAME AS FOR SINGLE CASED WELL.

GROUT.

PIPE TO PLACE ADDITIONAL GRAVEL.

BOTTOM OF GROUT

INSIDE CASING

OUTSIDE CASING

2"-4"

¼" WELL-ROUNDED AND WASHED GRAVEL.

SCREEN

DEPTH REQUIRED TO PREVENT SURFACE POLLUTION ENTERING GRAVEL.

DEPTH REQUIRED TO REACH WATERBEARING STRATA.

DOUBLE CASED GRAVEL WALL WELL

NOTE:

FOR USE WHERE WATER-BEARING STRATUM IS OF UNIFORM FINE SAND WHICH IS SUBJECT TO FLOWING AT THE VELOCITY THE WATER ENTERS THE SCREEN. THE GRAVEL PROVIDES A MUCH LARGER AREA THAN THE SCREEN AND THE VELOCITY OF WATER AT THE OUTER LINE OF GRAVEL MAY BE REDUCED TO SUCH VELOCITY THAT THE SAND WILL NOT FLOW.

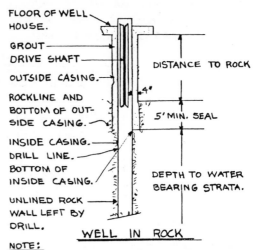

FLOOR OF WELL HOUSE.

GROUT

DRIVE SHAFT

OUTSIDE CASING.

ROCKLINE AND BOTTOM OF OUTSIDE CASING.

INSIDE CASING.

DRILL LINE.

BOTTOM OF INSIDE CASING.

UNLINED ROCK WALL LEFT BY DRILL.

DISTANCE TO ROCK

4"

5' MIN. SEAL

DEPTH TO WATER BEARING STRATA.

WELL IN ROCK

NOTE:

WELL HOUSE, PUMPING LEVEL, SCREEN AND OTHER DETAILS ARE SAME AS FOR SINGLE CASED WELL.

GENERAL NOTES:

1. WELL SITE SHOULD BE 200 TO 500 FEET FROM POSSIBLE SOURCE OF POLLUTION. LOCATE AT ELEVATED POINT IF POSSIBLE TO PREVENT FLOODING. FENCE IN WELL SITE.

2. WELL BUILDING SHOULD BE FIREPROOF, VENTILATED AND IN COLD CLIMATES, INSULATE THOROUGHLY AND PROVIDE HEAT. ELEVATE ABOVE GRADE TO PROVIDE DRAINAGE AWAY FROM BUILDING.

3. PUMPING LEVEL IS DETERMINED BY CONTINUOUS FLOW TEST AT REQUIRED WELL CAPACITY, AND THE MEASUREMENT OF THE RESULTING DRAWDOWN BELOW THE STATIC WATER TABLE.

4. THE TOP OF THE SCREEN SHOULD BE 50 FEET MINIMUM BELOW GRADE TO AVOID SURFACE POLLUTION UNLESS UNUSUALLY IMPERVIOUS EARTH (10 FEET OF COMPACT CLAY) OCCURS AT THE SURFACE, OR WELL IS FAR REMOVED FROM POSSIBLE SOURCES OF POLLUTION.

## FIGURE 2-7

Figures 2-7 and 2-8 are typical explanatory details of well systems to be given to designers and draftsmen. These can be used as details provided proper sizing and clarification is included and provided the installation can be related to a specific design.

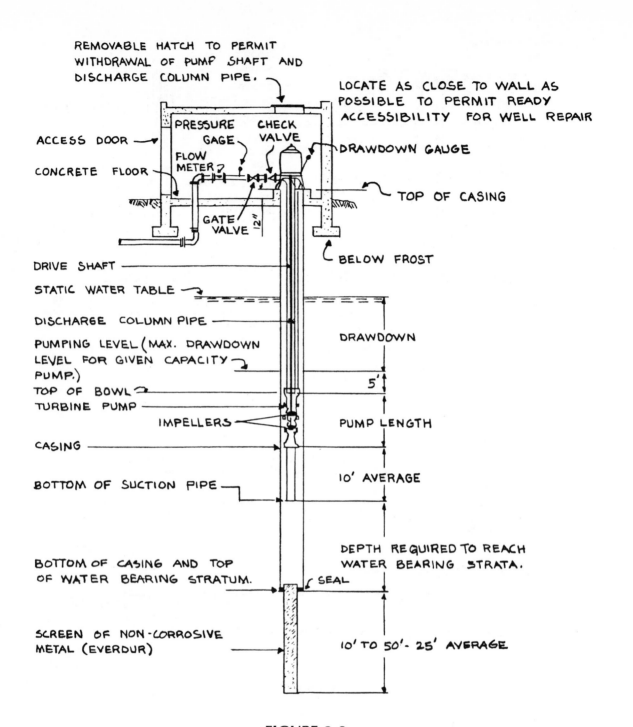

REMOVABLE HATCH TO PERMIT WITHDRAWAL OF PUMP SHAFT AND DISCHARGE COLUMN PIPE.

LOCATE AS CLOSE TO WALL AS POSSIBLE TO PERMIT READY ACCESSIBILITY FOR WELL REPAIR

ACCESS DOOR

PRESSURE GAGE

CHECK VALVE

FLOW METER

DRAWDOWN GAUGE

CONCRETE FLOOR

TOP OF CASING

GATE VALVE

12"

BELOW FROST

DRIVE SHAFT

STATIC WATER TABLE

DISCHARGE COLUMN PIPE

DRAWDOWN

PUMPING LEVEL (MAX. DRAWDOWN LEVEL FOR GIVEN CAPACITY PUMP.)

TOP OF BOWL

5'

TURBINE PUMP

IMPELLERS

PUMP LENGTH

CASING

BOTTOM OF SUCTION PIPE

10' AVERAGE

DEPTH REQUIRED TO REACH WATER BEARING STRATA.

BOTTOM OF CASING AND TOP OF WATER BEARING STRATUM.

SEAL

SCREEN OF NON-CORROSIVE METAL (EVERDUR)

10' TO 50' - 25' AVERAGE

## FIGURE 2-8

## SINGLE CASED WELL

FOR USE IN ORDINARY SANDY & GRAVELLY SOIL

188

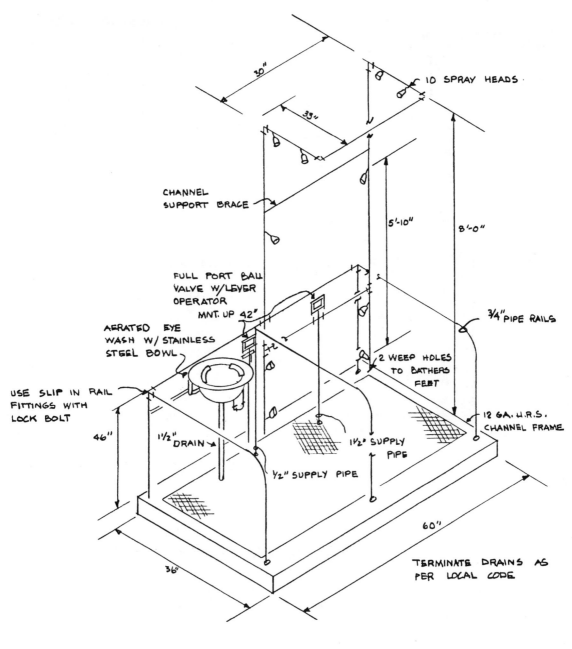

30"

33"

10 SPRAY HEADS

CHANNEL
SUPPORT BRACE

5'-10"

8'-0"

FULL PORT BALL
VALVE W/LEVER
OPERATOR
MNT. UP 42"

AERATED EYE
WASH W/ STAINLESS
STEEL BOWL

3/4" PIPE RAILS

2 WEEP HOLES
TO BATHERS
FEET

USE SLIP IN RAIL
FITTINGS WITH
LOCK BOLT

12 GA. H.R.S.
CHANNEL FRAME

46"

1 1/2" DRAIN

1 1/2" SUPPLY
PIPE

1/2" SUPPLY PIPE

60"

36"

TERMINATE DRAINS AS
PER LOCAL CODE

## FIGURE 2-9

MULTIPLE SPRAY SHOWER & EYE WASH DETAIL

Figure 2-9 is a typical pipe stand arrangement for a shower and eye wash
in an industrial boiler plant or other application requiring this type
installation. It is three dimensional with dimensions for certain of the
heads, which may be adjusted to actual conditions or, if space permits,
can be used as shown.

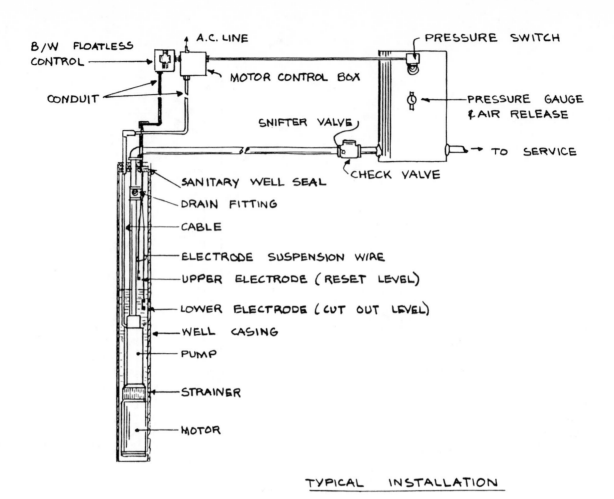

B/W FLOATLESS CONTROL

A.C. LINE

PRESSURE SWITCH

MOTOR CONTROL BOX

CONDUIT

PRESSURE GAUGE & AIR RELEASE

SNIFTER VALVE

TO SERVICE

CHECK VALVE

SANITARY WELL SEAL

DRAIN FITTING

CABLE

ELECTRODE SUSPENSION WIRE

UPPER ELECTRODE (RESET LEVEL)

LOWER ELECTRODE (CUT OUT LEVEL)

WELL CASING

PUMP

STRAINER

MOTOR

TYPICAL INSTALLATION

**FIGURE 2-10**

SUBMERSIBLE PUMP CONTROL

Figure 2-10 indicates the typical installation of a submersible pump control and is fairly clear and obvious in its installation. It is generally used as an explanatory detail, where required.

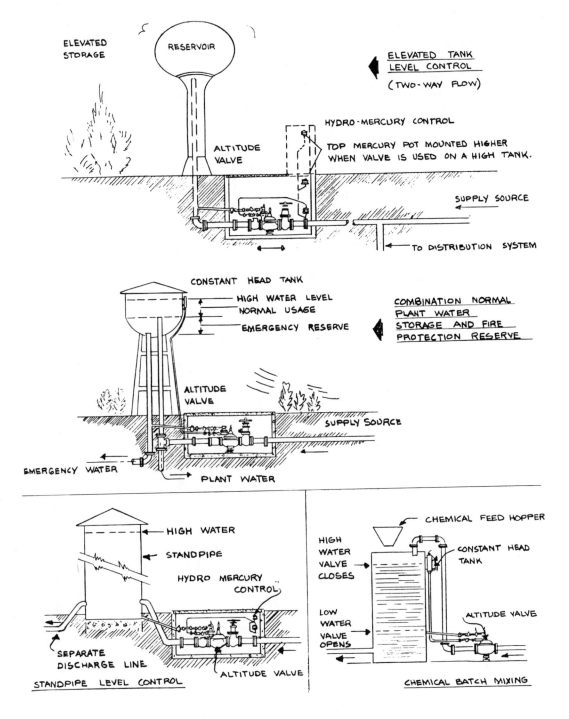

**FIGURE 2-11**

TYPICAL APPLICATIONS

Figure 2-11 is an explanatory drawing that can be given to designers and draftsmen. If used in an actual installation additional sizing and dimensioning would be required. The chemical treatment is simplified for clarification and explanatory purposes.

191

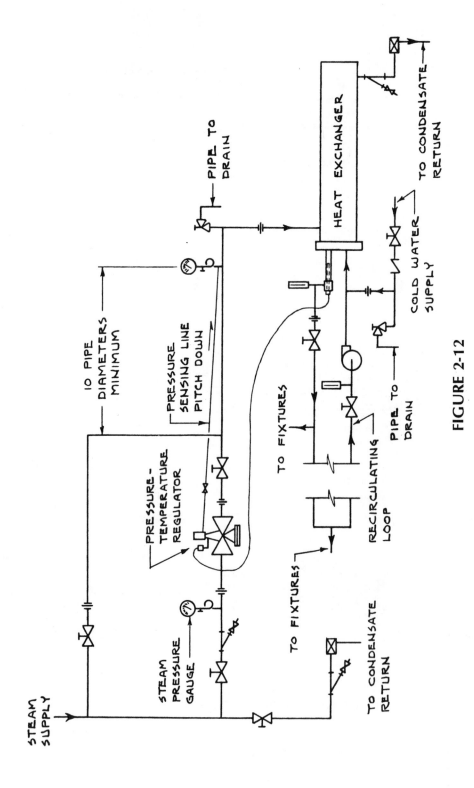

**FIGURE 2-12**

DOMESTIC HOT WATER FROM STEAM

Figure 2-12 is a standard detail for securing domestic hot water from a steam heated heat exchanger. This installation does not show some of the unions and pipe connections which perhaps should be installed. Generally, with minor adjustments this detail could be used as is.

192

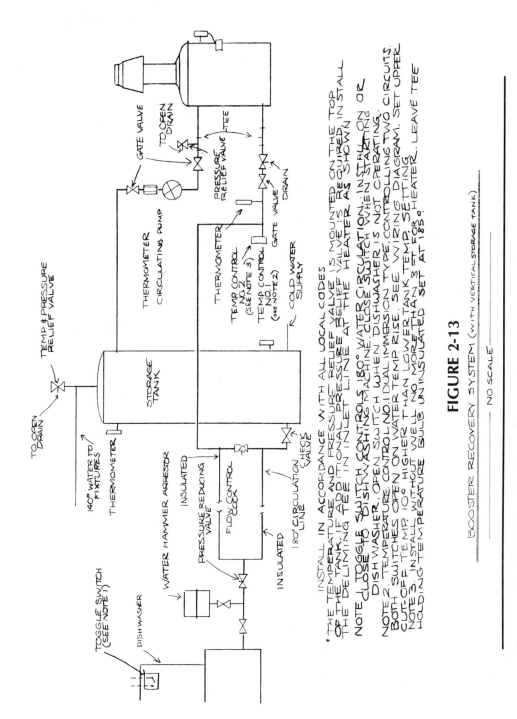

**FIGURE 2-13**

BOOSTER RECOVERY SYSTEM (WITH VERTICAL STORAGE TANK)

——— NO SCALE ———

Figure 2-13 is a relatively standard arrangement of a gas fired booster recovery system. It enables the system to provide 180° water to a dishwasher as well as 140° water to the system while at the same time allowing for a balance on the line in the form of the storage tank. The storage tank, of course, requires careful sizing by the engineer.

**193**

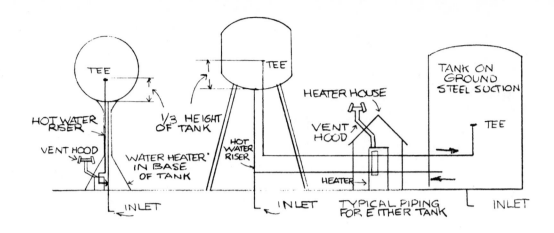

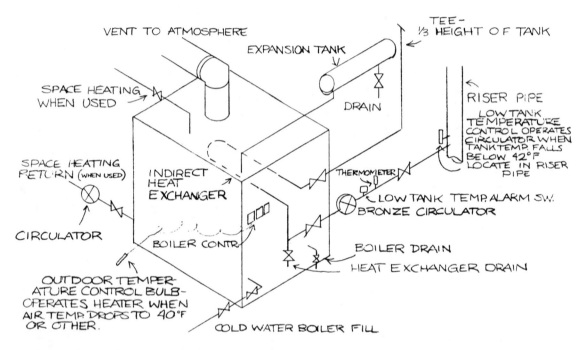

NOTE: ABOVE DETAILS IS FOR HEATER EQUIPPED WITH ONE HEAT EXCHANGER. HEATER CL8WT THRU CL210WT. L SERIES HEATERS HAVE TWO HEAT EXCHANGERS CONNECTED IN PARALLEL, OTHERWISE PIPING OF SYSTEMS REMAINS THE SAME.

**FIGURE 2-14**

TYPICAL BOILER PIPING
——— NO SCALE ———

Figure 2-14 is the typical boiler piping and generalized arrangement of providing heated water to a tank on the ground or an elevated tank. It is not intended to be traced as is, but to form the basis of a system carefully tailored to the application requirements.

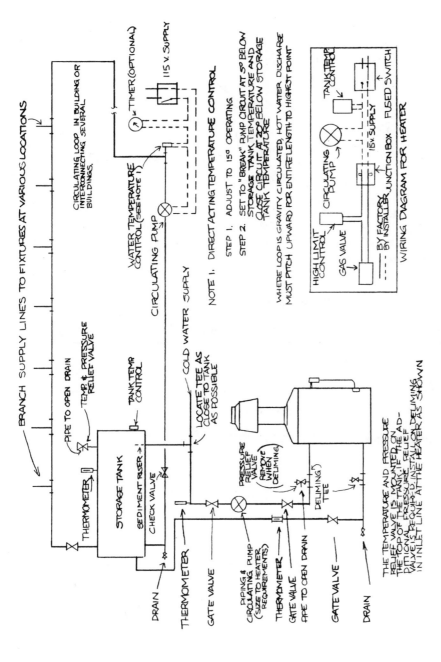

BRANCH SUPPLY LINES TO FIXTURES AT VARIOUS LOCATIONS

CIRCULATING LOOP IN BUILDING OR INTERCONNECTING SEVERAL BUILDINGS.

TIMER (OPTIONAL)

115 V. SUPPLY

WATER TEMPERATURE CONTROL (SEE NOTE 1)

CIRCULATING PUMP

COLD WATER SUPPLY

PIPE TO OPEN DRAIN

TEMP. & PRESSURE RELIEF VALVE

TANK TEMP. CONTROL

THERMOMETER

STORAGE TANK

SEDIMENT RISER

CHECK VALVE

DRAIN

THERMOMETER

GATE VALVE

PIPING & CIRCULATING PUMP (SIZE TO HEATER REQUIREMENTS)

LOCATE TEE AS CLOSE TO TANK AS POSSIBLE

PRESSURE RELIEF VALVE (REMOVE WHEN DELIMING)

DELIMING TEE

THERMOMETER

GATE VALVE

PIPE TO OPEN DRAIN

GATE VALVE

DRAIN

THE TEMPERATURE AND PRESSURE RELIEF VALVE IS MOUNTED ON THE TOP OF THE TANK. IF THE ADDITIONAL PRESSURE RELIEF VALVE IS REQUIRED, INSTALL ON DELIMING IN INLET LINE AT THE HEATER AS SHOWN

NOTE 1.   DIRECT ACTING TEMPERATURE CONTROL

STEP 1.   ADJUST TO 15° OPERATING

STEP 2.   SET TO "BREAK" PUMP CIRCUIT AT 5° BELOW STORAGE TANK TEMPERATURE AND CLOSE CIRCUIT AT 20° BELOW STORAGE TANK TEMPERATURE

WHERE LOOP IS GRAVITY CIRCULATED, HOT WATER DISCHARGE MUST PITCH UPWARD FOR ENTIRE LENGTH TO HIGHEST POINT

TANK TEMP CONTROL

CIRCULATING PUMP

115v. SUPPLY

FUSED SWITCH

HIGH LIMIT CONTROL

GAS VALVE

JUNCTION BOX

—— BY FACTORY
----- BY INSTALLER

WIRING DIAGRAM FOR HEATER

### FIGURE 2-15

RECOVERY SYSTEM WITH BUILDING CIRCULATING LOOP

— NO SCALE —

Figure 2-15 is another standard booster recovery system with a storage tank and circulating loop. Generally, it may be used as a schematic when this particular type of domestic water heating system is part of the contract.

195

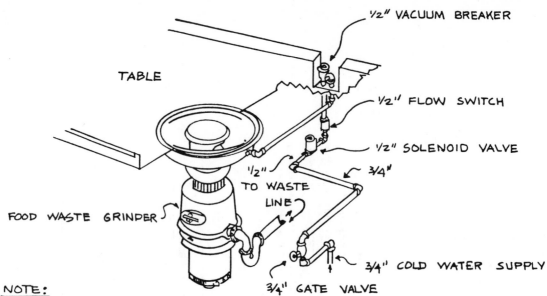

TABLE

½" VACUUM BREAKER

½" FLOW SWITCH

½" SOLENOID VALVE

3/4"

½"
TO WASTE
LINE

FOOD WASTE GRINDER

3/4" COLD WATER SUPPLY

3/4" GATE VALVE

NOTE:
ALL FOOD WASTE GRINDERS MUST
HAVE WATER TO OPERATE. WATER
SUPPLY MAY COME FROM FAUCET
ABOVE OR FROM DISCHARGE OF
DISWASHING MACHINE PRE-WASH.

**FIGURE 2-16**

TYPICAL PIPING & CONTROLS FOR FOOD WASTE
GRINDER

Figure 2-16 is a typical layout of piping and control lines for a food waste
grinder. It is one of those details that is relatively simple, but including it
on the plans will avoid arguments later.

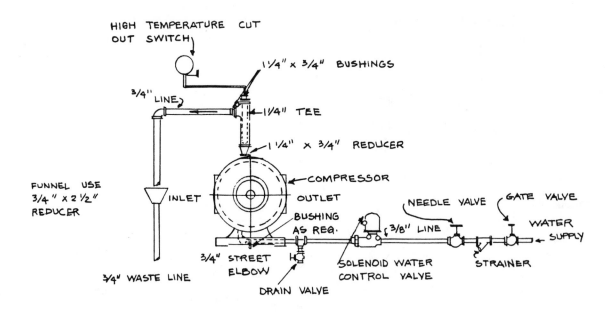

**FIGURE 2-17**

COMPRESSOR & VACUUM PUMP COOLING SYSTEMS

Figures 2-17 through 2-20 illustrate various arrangements of compressor and vacuum cooling systems. Note that there are slight differences in the way the control is used where a storage tank might be required or where a specialized system for compressors in stored air systems is required.

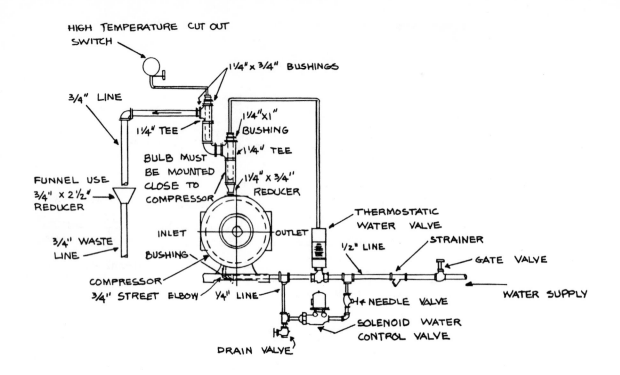

**FIGURE 2-18**

COMPRESSOR & VACUUM PUMP COOLING SYSTEMS

198

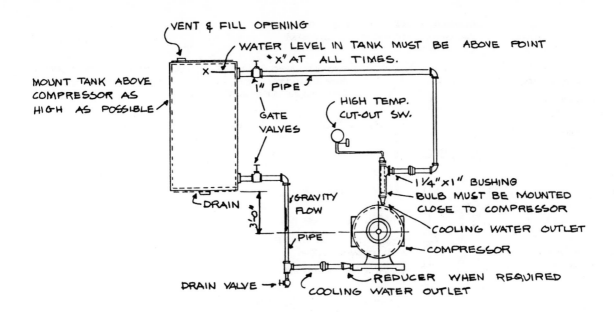

VENT & FILL OPENING

WATER LEVEL IN TANK MUST BE ABOVE POINT "X" AT ALL TIMES.

MOUNT TANK ABOVE COMPRESSOR AS HIGH AS POSSIBLE

X

1" PIPE

GATE VALVES

HIGH TEMP. CUT-OUT SW.

DRAIN

GRAVITY FLOW

3'-0"

PIPE

1¼"x1" BUSHING

BULB MUST BE MOUNTED CLOSE TO COMPRESSOR

COOLING WATER OUTLET

COMPRESSOR

DRAIN VALVE

COOLING WATER OUTLET

REDUCER WHEN REQUIRED

**FIGURE 2-19**

COMPRESSOR & VACUUM PUMP COOLING SYSTEMS

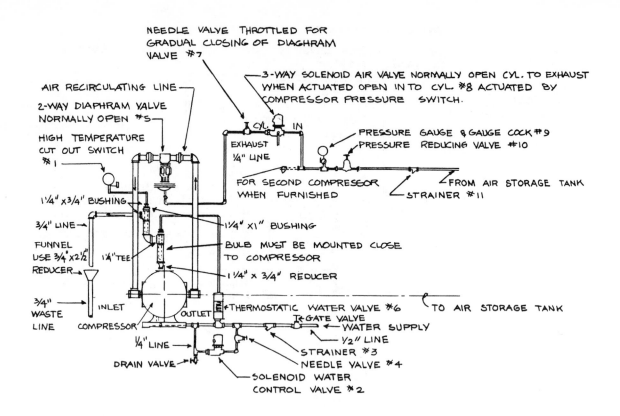

NEEDLE VALVE THROTTLED FOR GRADUAL CLOSING OF DIAGHRAM VALVE #7

3-WAY SOLENOID AIR VALVE NORMALLY OPEN CYL. TO EXHAUST WHEN ACTUATED OPEN INTO CYL #8 ACTUATED BY COMPRESSOR PRESSURE SWITCH.

AIR RECIRCULATING LINE

2-WAY DIAPHRAM VALVE NORMALLY OPEN #5

HIGH TEMPERATURE CUT OUT SWITCH #1

CYL. IN

EXHAUST ¼" LINE

PRESSURE GAUGE & GAUGE COCK #9
PRESSURE REDUCING VALVE #10

1¼" x ¾" BUSHING

FOR SECOND COMPRESSOR WHEN FURNISHED

FROM AIR STORAGE TANK

STRAINER #11

¾" LINE

1¼" x 1" BUSHING

FUNNEL USE ¾" x 2½" REDUCER

¼" TEE

BULB MUST BE MOUNTED CLOSE TO COMPRESSOR

1¼" x ¾" REDUCER

¾" WASTE LINE

INLET

COMPRESSOR

OUTLET

THERMOSTATIC WATER VALVE #6

TO AIR STORAGE TANK

GATE VALVE
WATER SUPPLY
½" LINE
STRAINER #3
NEEDLE VALVE #4

¼" LINE

DRAIN VALVE

SOLENOID WATER CONTROL VALVE #2

**FIGURE 2-20**

COMPRESSOR & VACUUM PUMP COOLING SYSTEMS

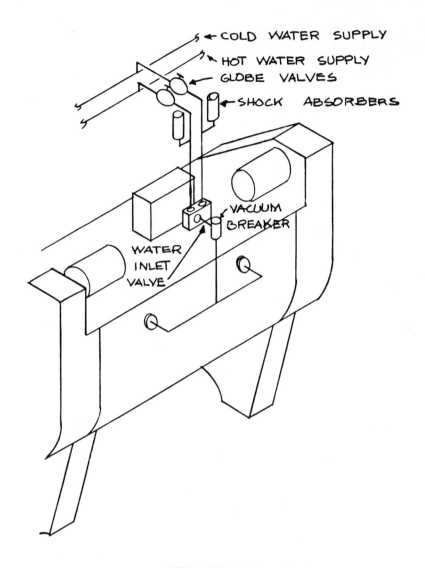

COLD WATER SUPPLY
HOT WATER SUPPLY
GLOBE VALVES
SHOCK ABSORBERS
VACUUM BREAKER
WATER INLET VALVE

**FIGURE 2-21**

TYPICAL WATER SUPPLY PIPING ARRANGEMENT FOR HOSPITAL WASHING MACHINES

Figure 2-21 is a relatively simple detail that emerged as a result of questions that arose with hospital washing machines and their connections. This detail is an excellent clarification drawing. If the design plans are at a sufficiently large scale, the elements of the detail could be shown on the plans.

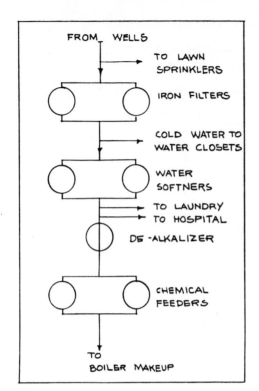

WELL WATER TREATMENT SYSTEM

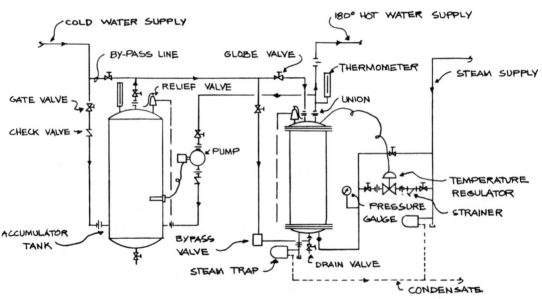

**FIGURE 2-22**

PIPING FOR ACCUMULATOR & CONTROLLED HEAT TRANSFER
RATE HOT WATER HEATER

Figure 2-22 is an illustration of two schematic systems—one for accumulator and control heat transfer and one for well water treatment—which are commonly used, when required, as explanatory type details.

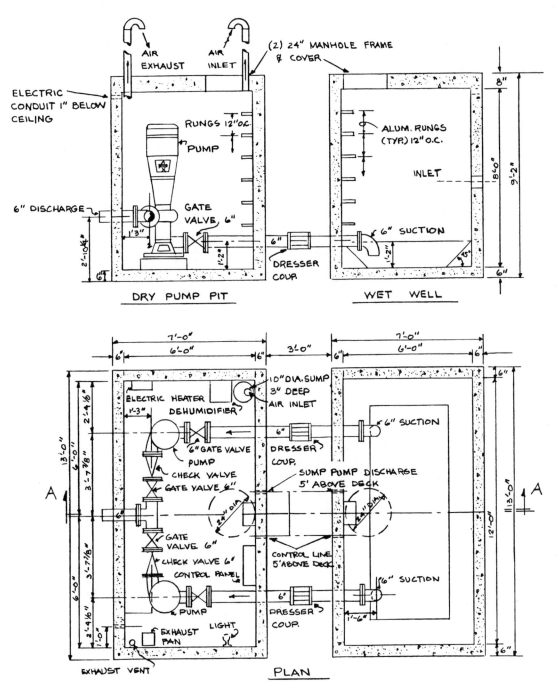

**FIGURE 2-23**

LIFT STATION WITH WET WELL

Figure 2-23 is a typical piping of a lift station. The most common
mistake in this detail is to forget the electric heater and the dehumidifier,
as well as the exhaust fan. If the sizing given for guidance happens to fit a
particular pump and system selection, the detail could be used as is.

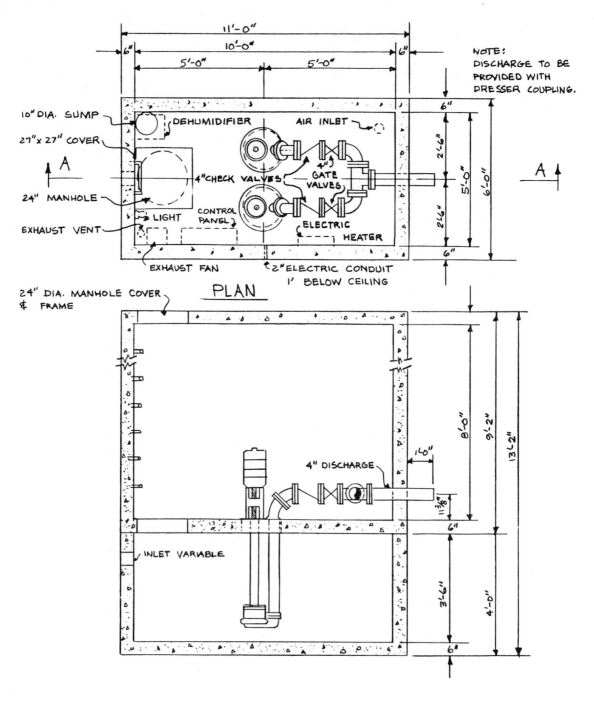

**FIGURE 2-24**

SUBMERGED TYPE · WET PIT SEWAGE
LIFT STATION

N.T.S.

Figure 2-24 is another version of the detail shown on Figure 2-23, but
instead of a wet well there is a wet pit.

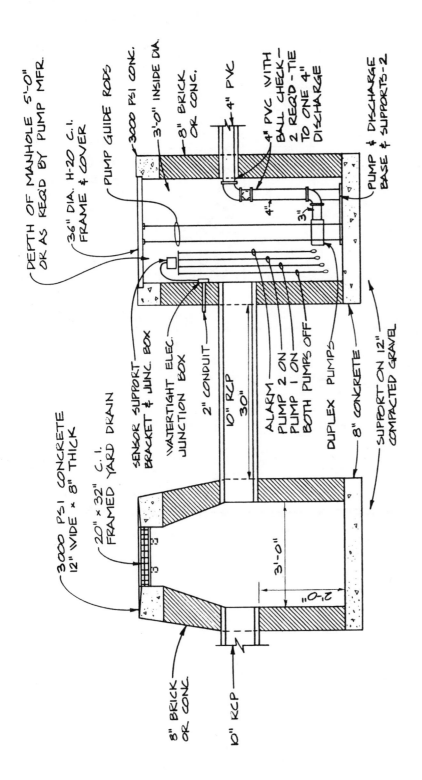

**FIGURE 2-25**

CATCH BASIN & LIFT PUMP STATION

NO SCALE

Figure 2-25 was specifically designed for an outdoor installation, for
example, in a parking lot or similar area, where it is necessary to lift the
storm water because of the pitch in the storm drainage system. If the area
is subject to heavy loads, an engineered structural solution of the manhole construction will be required.

205

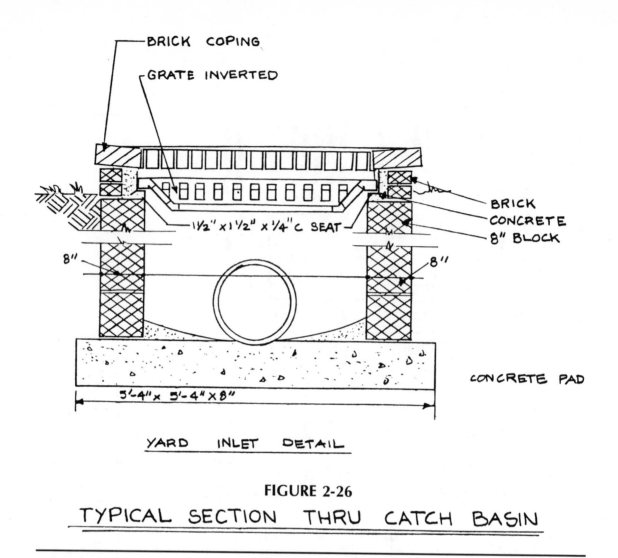

BRICK COPING

GRATE INVERTED

BRICK
CONCRETE
8" BLOCK

1½" x 1½" x ¼" C SEAT

8"

8"

CONCRETE PAD

5'-4" x 5'-4" x 8"

YARD INLET DETAIL

**FIGURE 2-26**

TYPICAL SECTION THRU CATCH BASIN

Figures 2-26 and 2-27 are fairly standard details showing a section through a catch basin. All construction should be checked against wheel load requirements.

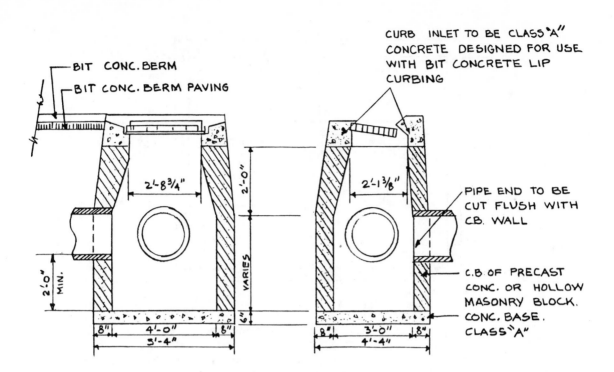

**FIGURE 2-27**

TYPICAL SECTION THRU PRECAST CATCH BASIN

207

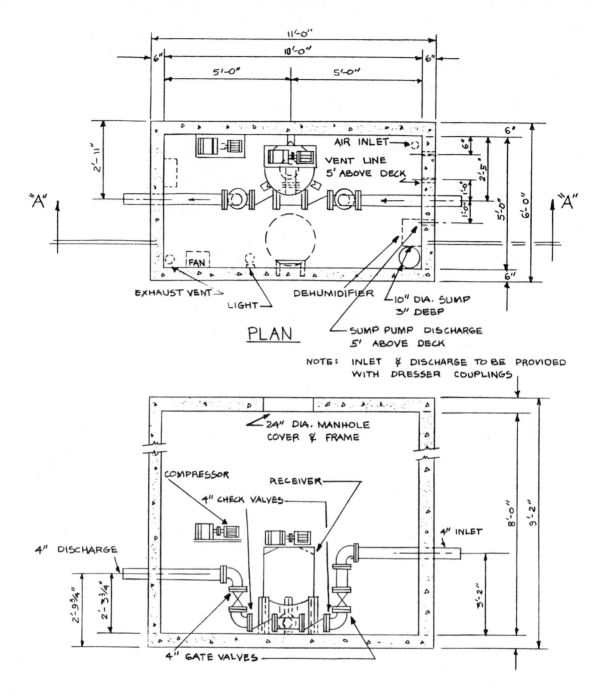

**FIGURE 2-28**

PNEUMATIC SEWAGE EJECTOR STATION

N.T.S.

Figure 2-28 is a standard detail of a pneumatic sewage ejector station. Note carefully the difference between inlet and discharge. The actual structural construction and reinforcement will have to be checked against the actual loads imposed. Sizing is for general guidance and may require adjustment to fit actual conditions.

208

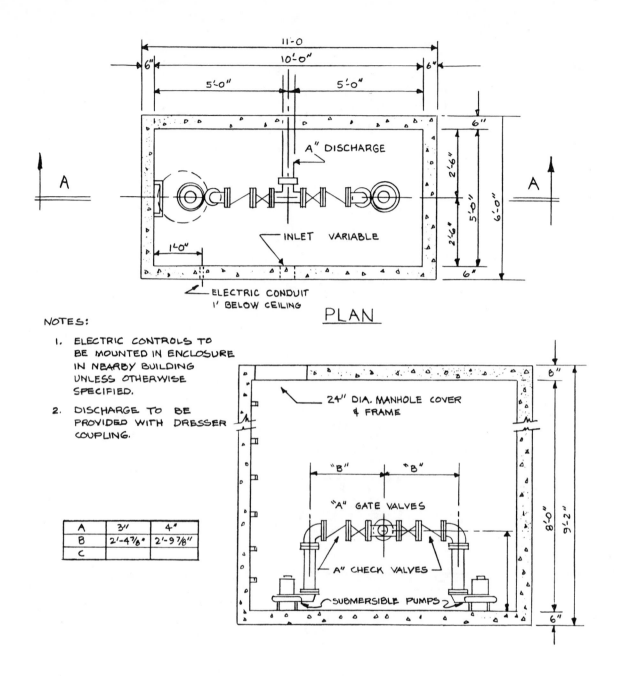

NOTES:

1. ELECTRIC CONTROLS TO BE MOUNTED IN ENCLOSURE IN NEARBY BUILDING UNLESS OTHERWISE SPECIFIED.

2. DISCHARGE TO BE PROVIDED WITH DRESSER COUPLING.

| A | 3" | 4" |
|---|---|---|
| B | 2'-4⅞" | 2'-9⅞" |
| C | | |

**FIGURE 2-29**

SUBMERSIBLE WET PIT SEWAGE LIFT STATION

N.T.S.

(RECOMMENDED FOR SEPTIC TANK EFFLUENT USE ONLY.)

Figure 2-29 is a typical piping for a sewage lift station as applied to a septic tank. There are no normal, common septic tank and field details in this book, since they are fairly common and can be obtained from many sources.

**209**

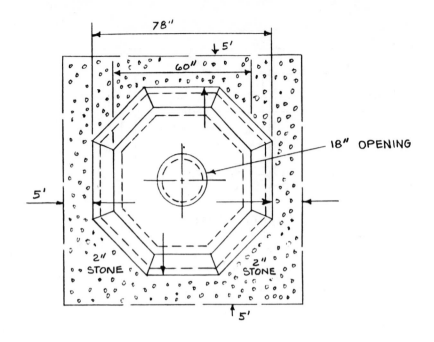

78"

↓ 5'

60"

18" OPENING

5'

2" STONE          2" STONE

5'

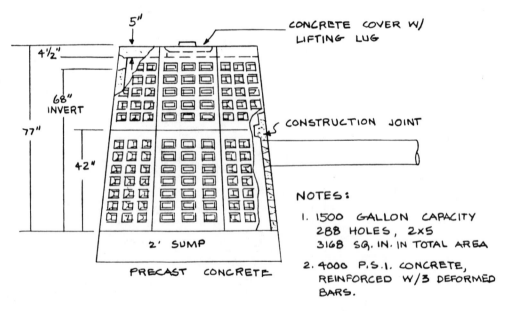

5"

CONCRETE COVER W/
LIFTING LUG

4½"

68" INVERT

77"

42"

CONSTRUCTION JOINT

2' SUMP

PRECAST CONCRETE

NOTES:

1. 1500 GALLON CAPACITY
   288 HOLES, 2x5
   3168 SQ. IN. IN TOTAL AREA

2. 4000 P.S.I. CONCRETE,
   REINFORCED W/3 DEFORMED
   BARS.

**FIGURE 2-30**

TYPICAL LEACHING GALLEY

NO SCALE

Figure 2-30, however, is a specialized situation where a septic field is not feasible. It is a leaching galley, which has to be carefully checked against local and state sanitary codes. The one shown has a 1500-gal capacity. The sizing would have to be varied and the number of gallies varied, depending on the load of effluent in the project.

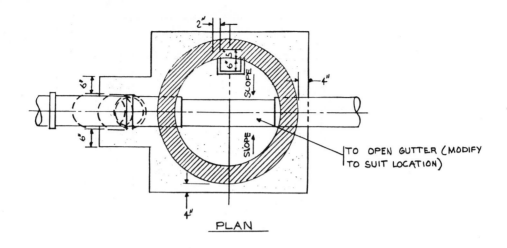

PLAN

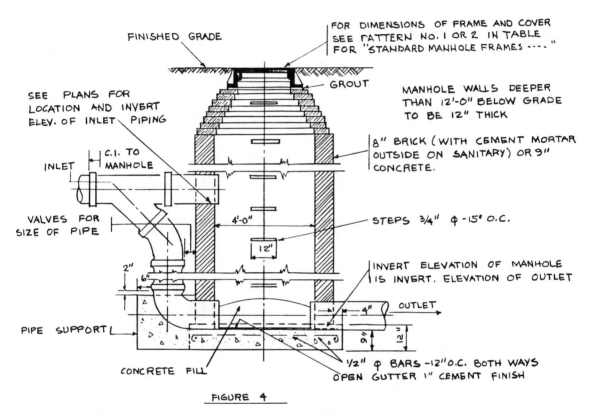

FINISHED GRADE

FOR DIMENSIONS OF FRAME AND COVER
SEE PATTERN NO. 1 OR 2 IN TABLE
FOR "STANDARD MANHOLE FRAMES ----"

GROUT

SEE PLANS FOR
LOCATION AND INVERT
ELEV. OF INLET PIPING

MANHOLE WALLS DEEPER
THAN 12'-0" BELOW GRADE
TO BE 12" THICK

C.I. TO
MANHOLE

INLET

8" BRICK (WITH CEMENT MORTAR
OUTSIDE ON SANITARY) OR 9"
CONCRETE.

VALVES FOR
SIZE OF PIPE

4'-0"

STEPS 3/4" φ -15' O.C.

12"

INVERT ELEVATION OF MANHOLE
IS INVERT. ELEVATION OF OUTLET

2"

6"

4"

OUTLET

9"

12"

PIPE SUPPORT

CONCRETE FILL

1/2" φ BARS -12"O.C. BOTH WAYS
OPEN GUTTER 1" CEMENT FINISH

TO OPEN GUTTER (MODIFY
TO SUIT LOCATION)

SLOPE

SLOPE

2"

6'

6'

4"

4"

FIGURE 4

**FIGURE 2-31**

STANDARD DROP MANHOLE, SANITARY AND DRAINAGE

Figure 2-31 is the standard drop manhole detail, which is commonly used
and is normally constructed of brick or concrete. The type of wheel
loading and location of the manhole would determine the actual construction of the manhole as well as its frame and cover.

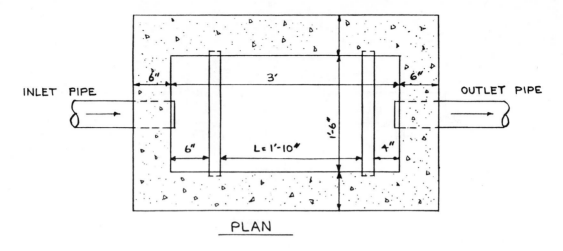

PLAN

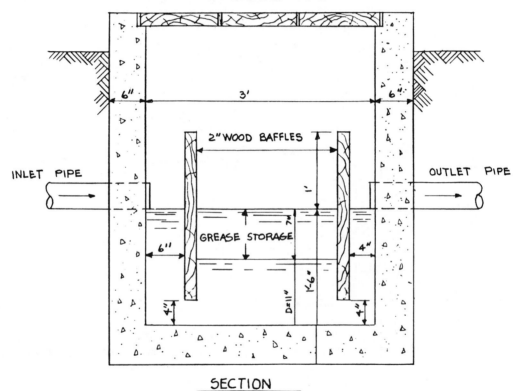

REMOVABLE COVER

INLET PIPE

OUTLET PIPE

2" WOOD BAFFLES

GREASE STORAGE

SECTION

SCALE 1"=1'-0"

**FIGURE 2-32**

DETAILS OF GREASE INTERCEPTOR

Figure 2-32 is another standard detail—in this case a grease interceptor which, although simple in arrangement, has proven to be very satisfactory. The sizing shown is representative. Actual sizing would vary according to the grease load conditions that exist.

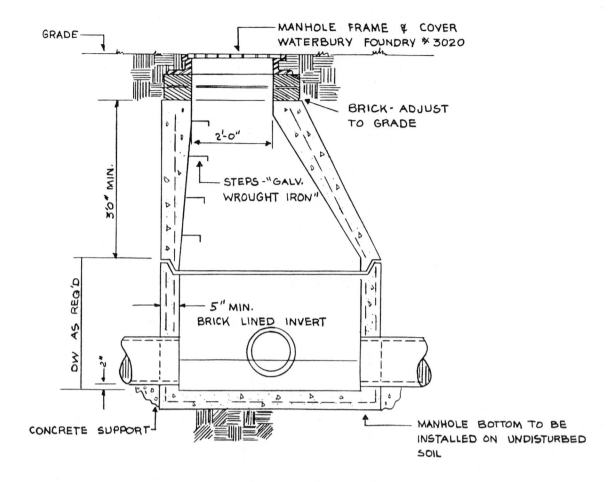

GRADE

MANHOLE FRAME & COVER
WATERBURY FOUNDRY # 3020

BRICK - ADJUST
TO GRADE

2'-0"

STEPS - "GALV.
WROUGHT IRON"

30" MIN.

5" MIN.
BRICK LINED INVERT

DW AS REQ'D

2"

CONCRETE SUPPORT

MANHOLE BOTTOM TO BE
INSTALLED ON UNDISTURBED
SOIL

NOTES:

1.  MANHOLE CONE & RISER SECTIONS TO CONFORM
    TO A.S.T.M. DESIGN C-478-61T.

2.  PROVIDE LIFTING HOLES ON ALL UNITS.

3.  ABSORPTION NOT TO EXCEED 8% AS PER
    A.S.T.M. C-76.

4.  FIELD BUILT MANHOLE WILL BE ACCEPTABLE-
    FURNISH SHOP DRAWINGS SHOWING CONST.
    DETAILS.

## FIGURE 2-33

## MANHOLE DETAIL

N.T.S.

Figures 2-33 through 2-36 are four standard manholes details. They are of various types and shapes, constructed either of concrete or brick and, in addition, a detail showing various types of manhole frames and covers is also given. Finally, in Figure 2-36 the capacity of the manhole, per foot of height, is listed. Wheel loading should be checked and reinforcement installed where required.

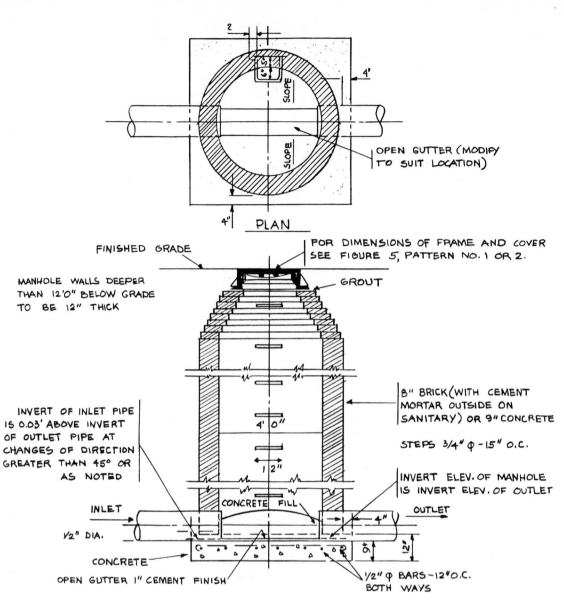

PLAN

FINISHED GRADE

FOR DIMENSIONS OF FRAME AND COVER SEE FIGURE 5, PATTERN NO. 1 OR 2.

MANHOLE WALLS DEEPER THAN 12'0" BELOW GRADE TO BE 12" THICK

GROUT

OPEN GUTTER (MODIFY TO SUIT LOCATION)

8" BRICK (WITH CEMENT MORTAR OUTSIDE ON SANITARY) OR 9" CONCRETE

STEPS 3/4" φ - 15" O.C.

INVERT OF INLET PIPE IS 0.03' ABOVE INVERT OF OUTLET PIPE AT CHANGES OF DIRECTION GREATER THAN 45° OR AS NOTED

4' 0"

12"

CONCRETE FILL

INVERT ELEV. OF MANHOLE IS INVERT ELEV. OF OUTLET

INLET

OUTLET

4"

1/2" DIA.

CONCRETE

9"

12"

OPEN GUTTER 1" CEMENT FINISH

1/2" φ BARS - 12" O.C. BOTH WAYS

SECTION

**FIGURE 2-34**

STANDARD MANHOLE, SANITARY & DRAINAGE

214

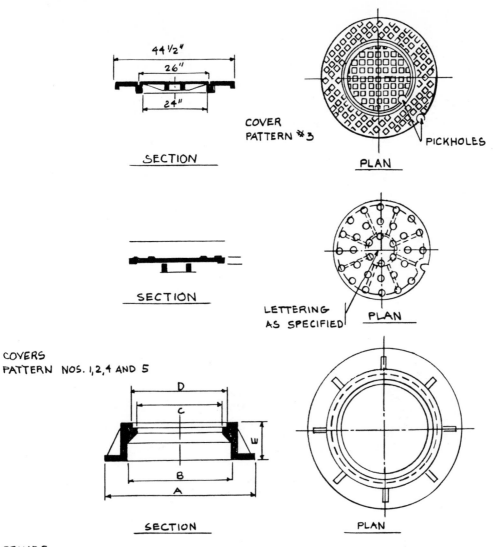

44½"

26"

24"

COVER
PATTERN #3

PICKHOLES

SECTION

PLAN

SECTION

LETTERING
AS SPECIFIED

PLAN

COVERS
PATTERN NOS. 1,2,4 AND 5

D

C

E

B

A

SECTION

PLAN

FRAMES
PATTERN NOS. 1,2,3,4,5

| PATTERN | WT., MIN. | A | B | C | D | E | F | G |
|---------|-----------|-----|-------|-------|-------|------|--------|-----|
| 1 | 440# | 36" | 26" | 20½" | 23¼" | 9" | 1-3/4" | 23" |
| 2 | 330# | 36" | 25½" | 20½" | 23¼" | 9" | 1-½" | 23" |
| 3 | 1075# | 56" | 46" | 42" | 44¾" | 10" | | |
| | 435# | 38" | 27" | 23" | 25¼" | 9" | 1½" | 25" |
| | 350# | 35" | 25½" | 23" | 25¼" | 7½" | 1½" | 24" |

FIGURE 5

NOTE: COVER FOR STORM MANHOLES TO BE SOLID OR PERFORATED
AS SPECIFIED
PATTERN NOS. 1 AND 4 FOR USE IN PAVED AREAS.
PATTERN NOS. 2 AND 5 FOR USE IN UNPAVED AREAS

**FIGURE 2-35**

STANDARD MANHOLE FRAMES & COVERS, PATTERNS #1,2,3,4 & 5

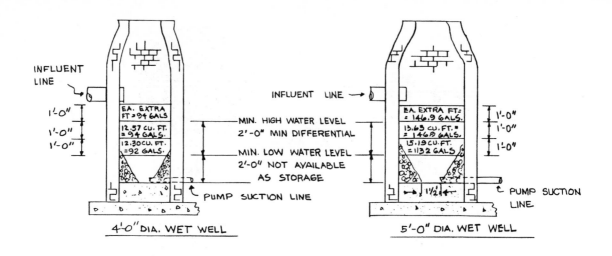

INFLUENT LINE

EA. EXTRA FT = 94 GALS.

12.57 CU. FT. = 94 GALS.

12.30 CU. FT. = 92 GALS.

1'-0"
1'-0"
1'-0"

MIN. HIGH WATER LEVEL
2'-0" MIN DIFFERENTIAL

MIN. LOW WATER LEVEL
2'-0" NOT AVAILABLE
AS STORAGE

PUMP SUCTION LINE

4'-0" DIA. WET WELL

INFLUENT LINE

EA. EXTRA FT. = 146.9 GALS.

13.63 CU. FT. = 146.9 GALS.

15.19 CU. FT. = 1132 GALS.

1'-0"
1'-0"
1'-0"

1½"

PUMP SUCTION LINE

5'-0" DIA. WET WELL

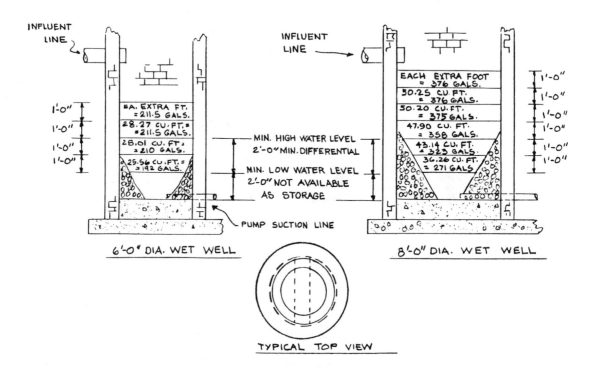

INFLUENT LINE

EA. EXTRA FT. = 211.5 GALS.

28.27 CU. FT. = 211.5 GALS.

28.01 CU. FT. = 210 GALS.

25.56 CU. FT. = 192 GALS.

1'-0"
1'-0"
1'-0"
1'-0"

MIN. HIGH WATER LEVEL
2'-0" MIN. DIFFERENTIAL

MIN. LOW WATER LEVEL
2'-0" NOT AVAILABLE
AS STORAGE

PUMP SUCTION LINE

6'-0" DIA. WET WELL

INFLUENT LINE

EACH EXTRA FOOT = 376 GALS.

50.25 CU. FT. = 376 GALS.

50.20 CU. FT. = 375 GALS.

47.90 CU. FT. = 358 GALS.

43.14 CU. FT. = 323 GALS.

36.26 CU. FT. = 271 GALS.

1'-0"
1'-0"
1'-0"
1'-0"
1'-0"
1'-0"

8'-0" DIA. WET WELL

TYPICAL TOP VIEW

## FIGURE 2-36

## TYPICAL WET WELL DESIGNS

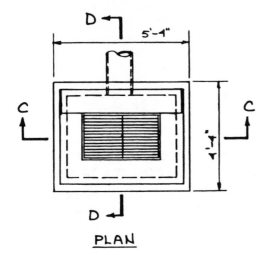

PLAN

WITH CONCRETE MASONRY CORBELLING WILL BE PERMITTED TO A MAX. OF 3"

CATCH BASIN WALL TO BE 12" THICK WHEN DEPTH OF MANHOLE IS GREATER THAN 10' (MASONRY ONLY)

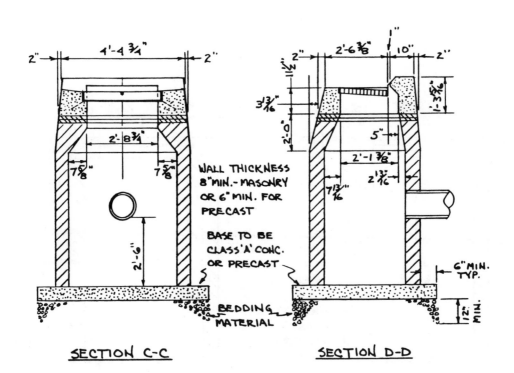

SECTION C-C

SECTION D-D

WALL THICKNESS 8" MIN.- MASONRY OR 6" MIN. FOR PRECAST

BASE TO BE CLASS 'A' CONC. OR PRECAST

BEDDING MATERIAL

FIGURE 2-37

CURB TYPE CATCH BASIN

NOT TO SCALE

Figures 2-37 and 2-38 are standard drawings of catch basins constructed of brick or concrete. These can be used as is. Wheel loading should be checked.

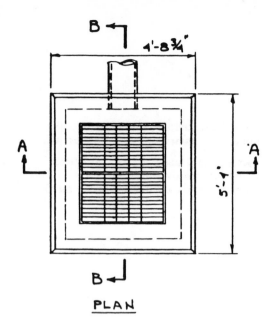

PLAN

WITH CONCRETE MASONRY CORBELLING WILL BE PERMITTED TO A MAX. OF 3"

CATCH BASIN WALLS TO BE 12" THICK WHEN DEPTH OF MANHOLE IS GREATER THAN 10' (MASONRY ONLY)

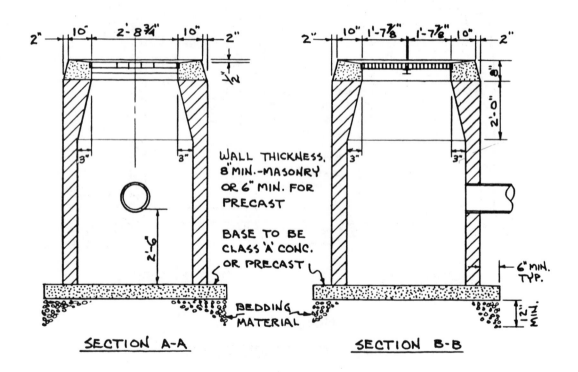

SECTION A-A

WALL THICKNESS, 8" MIN.-MASONRY OR 6" MIN. FOR PRECAST

BASE TO BE CLASS 'A' CONC. OR PRECAST

BEDDING MATERIAL

SECTION B-B

FIGURE 2-38

DOUBLE GRATE CATCH BASIN

NOT TO SCALE

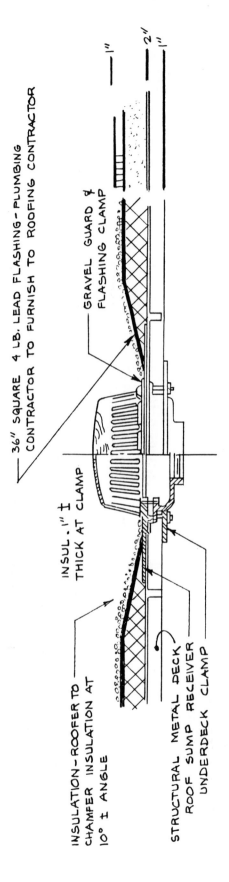

36" SQUARE 4 LB. LEAD FLASHING-PLUMBING
CONTRACTOR TO FURNISH TO ROOFING CONTRACTOR

GRAVEL GUARD &
FLASHING CLAMP

1"

2"
1"

INSUL. 1" ±
THICK AT CLAMP

INSULATION-ROOFER TO
CHAMFER INSULATION AT
10° ± ANGLE

STRUCTURAL METAL DECK
ROOF SUMP RECEIVER
UNDERDECK CLAMP

1. LOCATE AT LOW POINTS OF ROOF.

2. PIPE WITH EXPANSION JOINTS OR
   OFFSETS TO ALLOW FOR EXPANSION.

3. FURNISH SUMP RECEIVER TO GENERAL
   CONTRACTOR SO RECEIVER CAN BE
   ATTACHED TO THE DECK.

## FIGURE 2-39

ROOF   DRAIN   DETAIL

NO SCALE

Figure 2-39 is a standard roof drain detail. The type of roof drain actually used and its appearance may vary. The most important part about this detail is the notation on the flashing and the work that the roofer is to do. This point should be carefully related to all affected sections of the specifications.

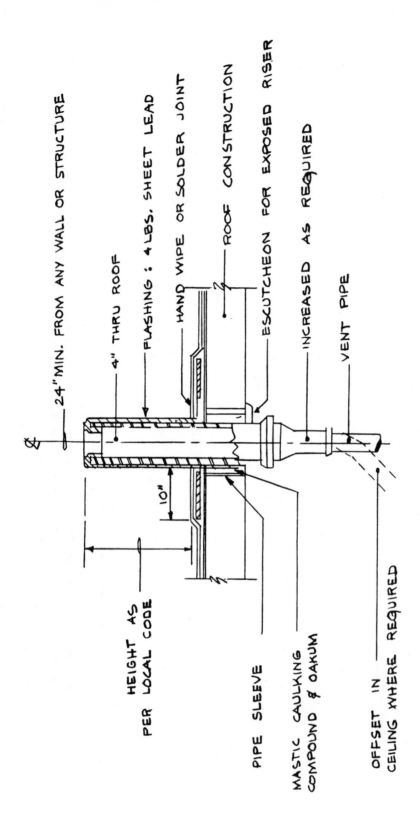

CL 24" MIN. FROM ANY WALL OR STRUCTURE

4" THRU ROOF

FLASHING : 4 LBS. SHEET LEAD

HAND WIPE OR SOLDER JOINT

ROOF CONSTRUCTION

ESCUTCHEON FOR EXPOSED RISER

INCREASED AS REQUIRED

VENT PIPE

HEIGHT AS PER LOCAL CODE

10"

PIPE SLEEVE

MASTIC CAULKING COMPOUND & OAKUM

OFFSET IN CEILING WHERE REQUIRED

**FIGURE 2-40**

VENT THRU ROOF DETAIL

NO SCALE

Figure 2-40 is a standard cast-iron vent detail. The vent, in this case, is shown as being increased from 2" to 4". Some localities in the country will allow a 3" vent, and this notation should be adjusted for the locality in which the engineer is working.

220

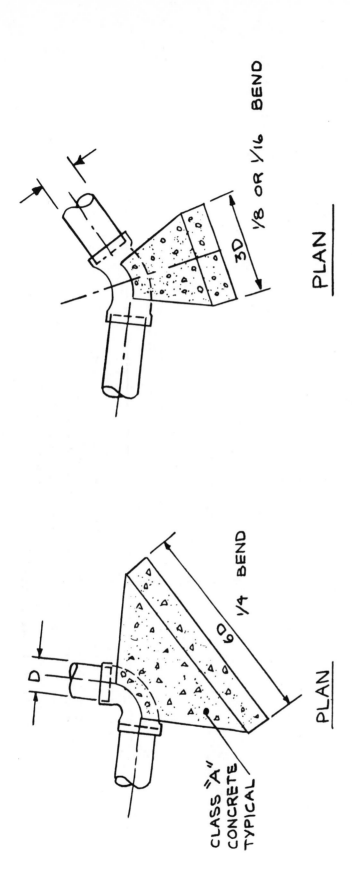

**FIGURE 2-41**

CONCRETE THRUST BLOCK

NO SCALE

Figure 2-41 is a relatively simple detail which requires calculations and sizing for proper dimensions, depending on the size of the pipe and the thrust that is involved. The general arrangement and relationship of the size of the thrust block is based on pipe diameter and is a common standard.

221

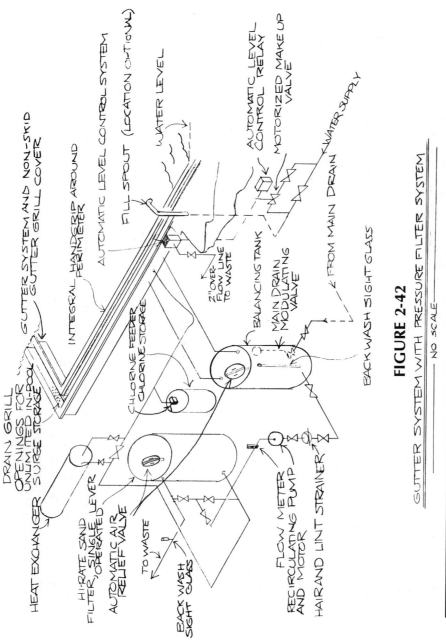

**FIGURE 2-42**

GUTTER SYSTEM WITH PRESSURE FILTER SYSTEM

— NO SCALE —

Figures 2-42 through 2-46 are typical swimming pool filter installations. They are schematic drawings and cover pressure filters, high rate sand filters (with the filter above and below the deck), and the standard YMCA three tank sand filters. Finally, the diatomite filter is also covered. These are not details as such, but should be used as design clarification and guidance material. As details they would require more careful fitting and relating to actual space conditions.

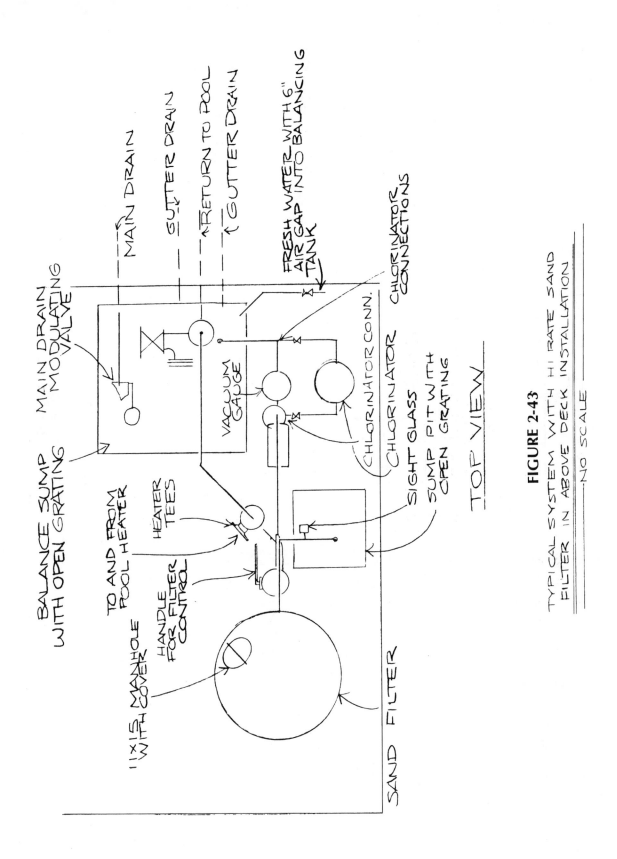

MAIN DRAIN

GUTTER DRAIN

RETURN TO POOL

GUTTER DRAIN

FRESH WATER WITH 6"
AIR GAP INTO BALANCING
TANK

CHLORINATOR
CONNECTIONS

MAIN DRAIN
MODULATING
VALVE

CHLORINATOR CONN.

CHLORINATOR

SUMP PIT WITH
OPEN GRATING

SIGHT GLASS

BALANCE SUMP
WITH OPEN GRATING

VACUUM GAUGE

TO AND FROM
POOL HEATER

HEATER
TEES

1'X15' MANHOLE
WITH COVER

HANDLE
FOR FILTER
CONTROL

SAND FILTER

**TOP VIEW**

TYPICAL SYSTEM WITH HI RATE SAND
FILTER IN ABOVE DECK INSTALLATION

NO SCALE

**FIGURE 2-43**

223

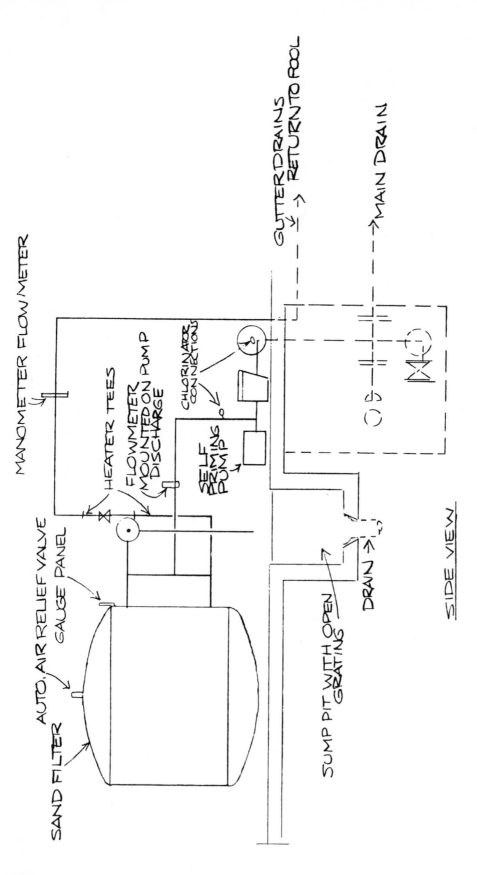

MANOMETER FLOW METER

HEATER TEES

FLOWMETER
(MOUNTED ON PUMP
DISCHARGE

CHLORINATOR
CONNECTIONS

SELF
PRIMING
PUMP

AUTO. AIR RELIEF VALVE
GAUGE PANEL

SAND FILTER

SUMP PIT WITH OPEN
GRATING

DRAIN →

GUTTER DRAINS
↓
→ RETURN TO POOL

→ MAIN DRAIN

SIDE VIEW

TYPICAL SYSTEM WITH HI RATE SAND FILTER IN ABOVE
DECK INSTALLATION
NO SCALE

**FIGURE 2-44**

224

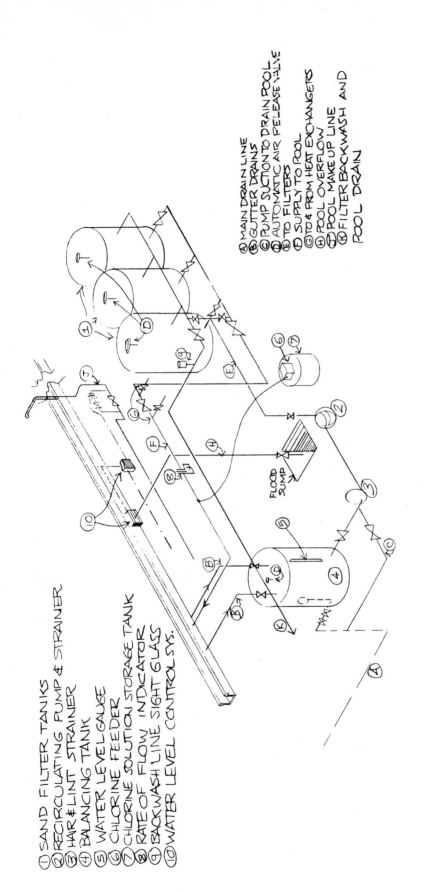

1. SAND FILTER TANKS
2. RECIRCULATING PUMP & STRAINER
3. HAIR & LINT STRAINER
4. BALANCING TANK
5. WATER LEVEL GAUGE
6. CHLORINE FEEDER
7. CHLORINE SOLUTION STORAGE TANK
8. RATE OF FLOW INDICATOR
9. BACKWASH LINE SIGHT GLASS
10. WATER LEVEL CONTROL SYS.

A. MAIN DRAIN LINE
B. GUTTER DRAINS
C. PUMP SUCTION TO DRAIN POOL
D. AUTOMATIC AIR RELEASE VALVE
E. TO FILTERS
F. SUPPLY TO POOL
G. TO & FROM HEAT EXCHANGERS
H. POOL OVERFLOW
J. POOL MAKE UP LINE
K. FILTER BACKWASH AND POOL DRAIN

FLOOD SUMP

**FIGURE 2-45**

GUTTER SYSTEM WITH YMCA
3 TANK SAND FILTER SYSTEM
NO SCALE

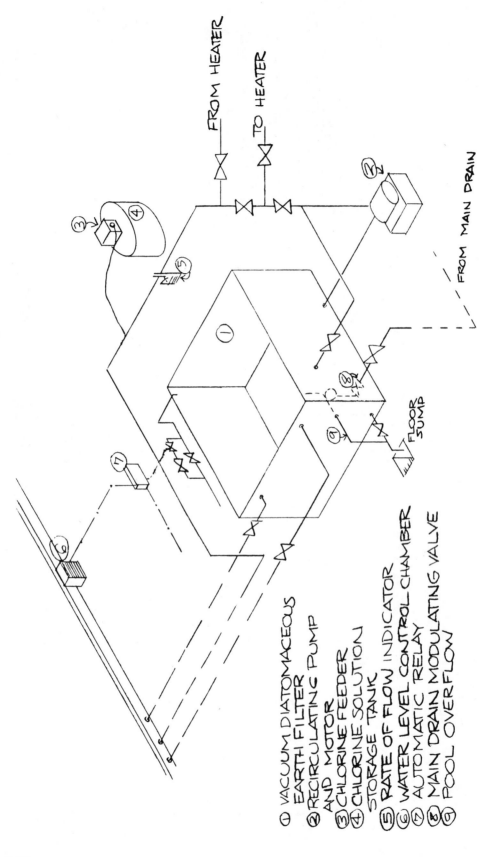

FROM HEATER

TO HEATER

FROM MAIN DRAIN

FLOOR SUMP

**FIGURE 2-46**

POOL SYSTEM WITH DIATOMITE FILTER

NO SCALE

① VACUUM DIATOMACEOUS
   EARTH FILTER
② RECIRCULATING PUMP
   AND MOTOR
③ CHLORINE FEEDER
④ CHLORINE SOLUTION
   STORAGE TANK
⑤ RATE OF FLOW INDICATOR
⑥ WATER LEVEL CONTROL CHAMBER
⑦ AUTOMATIC RELAY
⑧ MAIN DRAIN MODULATING VALVE
⑨ POOL OVERFLOW

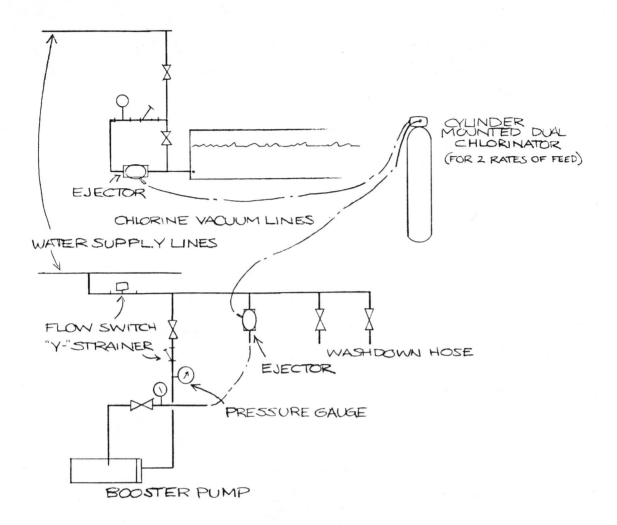

CYLINDER
MOUNTED DUAL
CHLORINATOR
(FOR 2 RATES OF FEED)

EJECTOR

CHLORINE VACUUM LINES

WATER SUPPLY LINES

FLOW SWITCH
"Y-"STRAINER

EJECTOR

WASHDOWN HOSE

PRESSURE GAUGE

BOOSTER PUMP

**FIGURE 2-47**

DUAL CHLORINATOR SYSTEM FOR APPLYING CHLORINE
TO TWO SYSTEMS REQUIRING DIFFERENT CHLORINE
CONCENTRATION

NO SCALE

Figures 2-47 through 2-50 are schematic details without all the fittings
necessarily shown. They are guide details for the injection of chlorine into
a system. This point is frequently misunderstood or not clearly shown on
the plans. This injection positioning is incorporated in a location appropriate on most plans that require chlorine application.

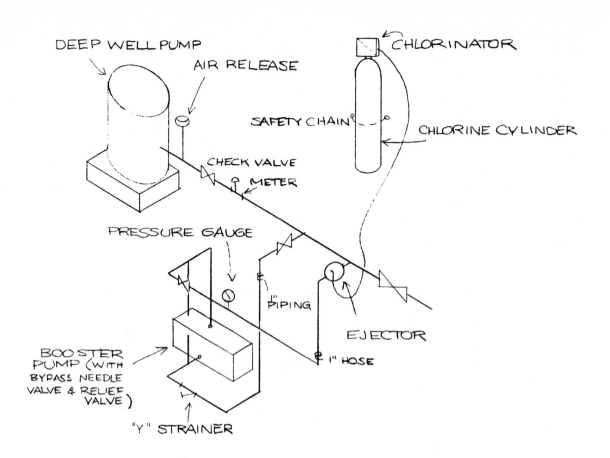

**FIGURE 2-48**

CHLORINATOR INSTALLATION USING
TURBINE TYPE BOOSTER PUMP

———— NO SCALE ————

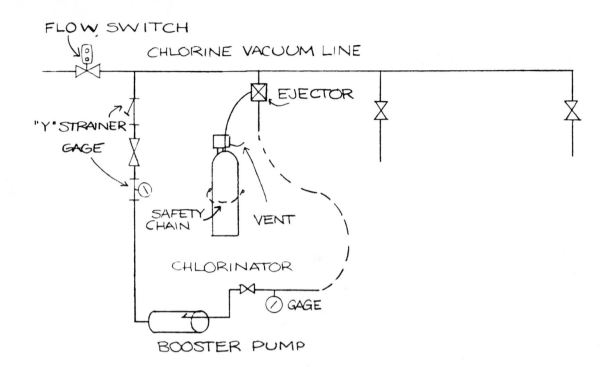

FLOW SWITCH

CHLORINE VACUUM LINE

EJECTOR

"Y" STRAINER
GAGE

SAFETY
CHAIN

VENT

CHLORINATOR

GAGE

BOOSTER PUMP

**FIGURE 2-49**

CHLORINATOR INSTALLATION FOR APPLYING
CHLORINE INTO PIPELINE USING A BOOSTER PUMP
—— NO SCALE ——

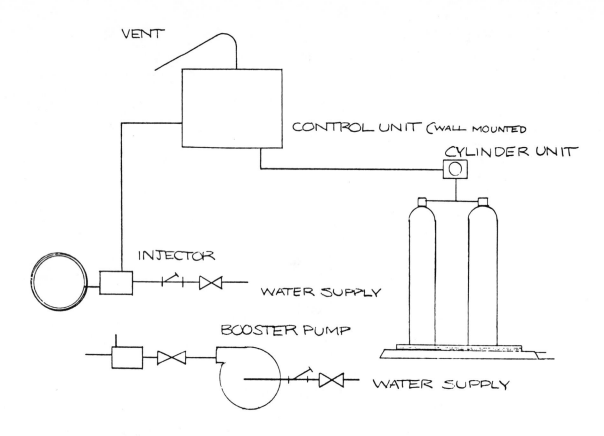

VENT

CONTROL UNIT (WALL MOUNTED

CYLINDER UNIT

INJECTOR

WATER SUPPLY

BOOSTER PUMP

WATER SUPPLY

**FIGURE 2-50**

TYPICAL WALL MOUNTED CHLORINATOR
FOR SWIMMING POOLS
NO SCALE

230

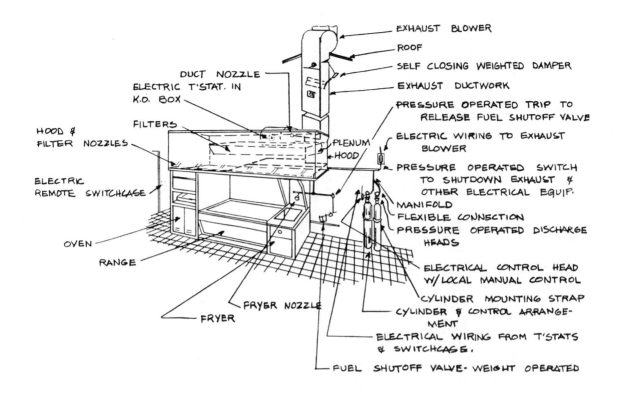

**FIGURE 2-51**

## CARBON DIOXIDE FIRE EXTINGUISHING SYSTEMS

PROTECTION FOR COMMERCIAL KITCHEN EQUIPMENT

Figures 2-51 through 2-54 are a series of details concerning a small system that is seldom used. But when it's needed, it's needed. In cases where the $CO^2$ fire extinguishing systems shown in these drawings are required, the simplest solution was to provide three dimensional drawings in lieu of cluttering the plan (which is usually already cluttered with other piping). These details can be put on as a separate plan and detail with a note on the regular plan referring to the detail for the $CO^2$ fire extinguishing system. There are any number of uses for $CO^2$. The ones shown here are for kitchens, air cleaners, and hoods on ovens.

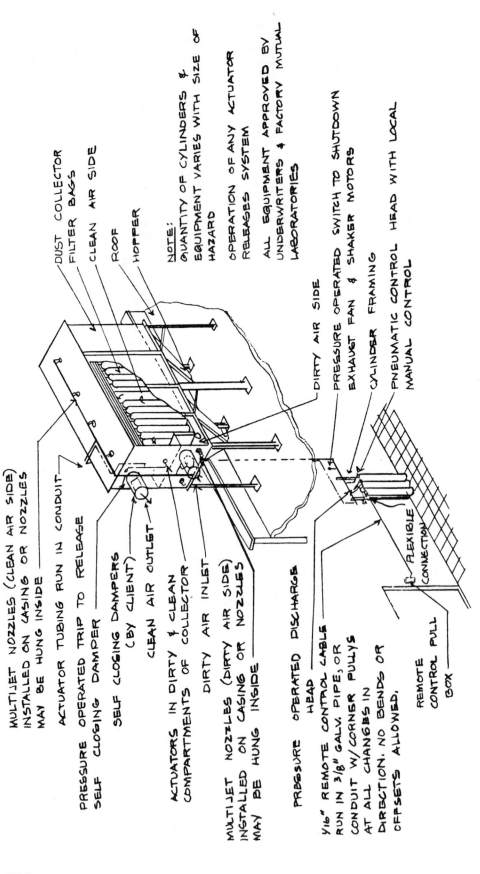

MULTIJET NOZZLES (CLEAN AIR SIDE) INSTALLED ON CASING OR NOZZLES MAY BE HUNG INSIDE

ACTUATOR TUBING RUN IN CONDUIT

PRESSURE OPERATED TRIP TO RELEASE SELF CLOSING DAMPER

SELF CLOSING DAMPERS (BY CLIENT)

CLEAN AIR OUTLET

ACTUATORS IN DIRTY & CLEAN COMPARTMENTS OF COLLECTOR

DIRTY AIR INLET

MULTIJET NOZZLES (DIRTY AIR SIDE) INSTALLED ON CASING OR NOZZLES MAY BE HUNG INSIDE

PRESSURE OPERATED DISCHARGE HEAD

1/16" REMOTE CONTROL CABLE RUN IN 3/8" GALV. PIPE, OR CONDUIT W/CORNER PULLYS AT ALL CHANGES IN DIRECTION. NO BENDS OR OFFSETS ALLOWED.

REMOTE CONTROL PULL BOX

DUST COLLECTOR FILTER BAGS

CLEAN AIR SIDE

ROOF

HOPPER

NOTE:
QUANTITY OF CYLINDERS & EQUIPMENT VARIES WITH SIZE OF HAZARD

OPERATION OF ANY ACTUATOR RELEASES SYSTEM

ALL EQUIPMENT APPROVED BY UNDERWRITERS & FACTORY MUTUAL LABORATORIES

DIRTY AIR SIDE

PRESSURE OPERATED SWITCH TO SHUTDOWN EXHAUST FAN & SHAKER MOTORS

CYLINDER FRAMING

PNEUMATIC CONTROL HEAD WITH LOCAL MANUAL CONTROL

FLEXIBLE CONNECTION

FIGURE 2-52

CARBON DIOXIDE FIRE EXTINGUISHING SYSTEMS

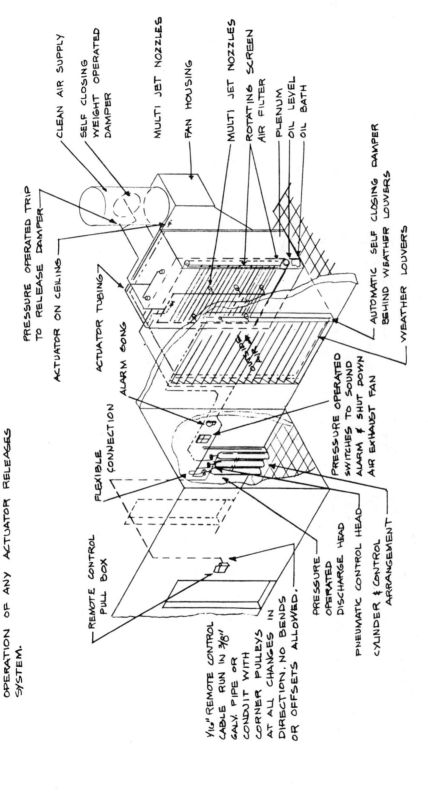

NOTE:

QUANTITY OF CYLINDERS & EQUIPMENT VARIES WITH SIZE OF HAZARD.

OPERATION OF ANY ACTUATOR RELEASES SYSTEM.

CLEAN AIR SUPPLY

SELF CLOSING WEIGHT OPERATED DAMPER

MULTI JET NOZZLES

FAN HOUSING

MULTI JET NOZZLES

ROTATING SCREEN AIR FILTER

PLENUM

OIL LEVEL

OIL BATH

PRESSURE OPERATED TRIP TO RELEASE DAMPER

ACTUATOR ON CEILING

ACTUATOR TUBING

ALARM GONG

FLEXIBLE CONNECTION

REMOTE CONTROL PULL BOX

1/16" REMOTE CONTROL CABLE RUN IN 3/8" GALV. PIPE OR CONDUIT WITH CORNER PULLEYS AT ALL CHANGES IN DIRECTION, NO BENDS OR OFFSETS ALLOWED.

PRESSURE OPERATED DISCHARGE HEAD

PNEUMATIC CONTROL HEAD

CYLINDER & CONTROL ARRANGEMENT

PRESSURE OPERATED SWITCHES TO SOUND ALARM & SHUT DOWN AIR EXHAUST FAN

AUTOMATIC SELF CLOSING DAMPER BEHIND WEATHER LOUVERS

WEATHER LOUVERS

FIGURE 2-53

CARBON DIOXIDE FIRE EXTINGUISHING SYSTEM

PROTECTION FOR OIL BATH AIR FILTERS

233

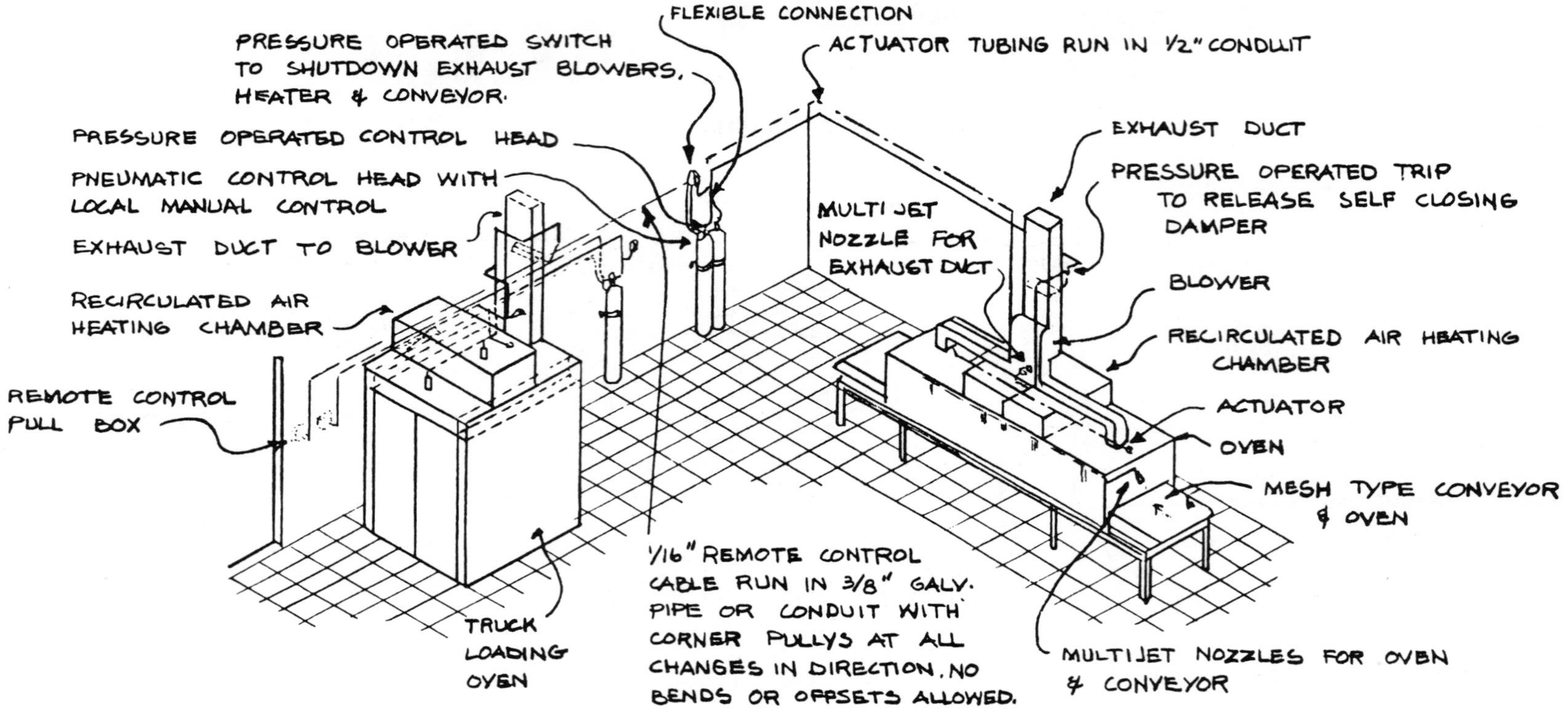

FLEXIBLE CONNECTION

ACTUATOR TUBING RUN IN 1/2" CONDUIT

EXHAUST DUCT

PRESSURE OPERATED TRIP TO RELEASE SELF CLOSING DAMPER

BLOWER

RECIRCULATED AIR HEATING CHAMBER

ACTUATOR

OVEN

MESH TYPE CONVEYOR & OVEN

MULTI JET NOZZLE FOR EXHAUST DUCT

MULTIJET NOZZLES FOR OVEN & CONVEYOR

PRESSURE OPERATED SWITCH TO SHUTDOWN EXHAUST BLOWERS, HEATER & CONVEYOR.

PRESSURE OPERATED CONTROL HEAD

PNEUMATIC CONTROL HEAD WITH LOCAL MANUAL CONTROL

EXHAUST DUCT TO BLOWER

RECIRCULATED AIR HEATING CHAMBER

REMOTE CONTROL PULL BOX

1/16" REMOTE CONTROL CABLE RUN IN 3/8" GALV. PIPE OR CONDUIT WITH CORNER PULLYS AT ALL CHANGES IN DIRECTION. NO BENDS OR OFFSETS ALLOWED.

TRUCK LOADING OVEN

**FIGURE 2-54**

CARBON DIOXIDE FIRE EXTINGUISHING SYSTEM

PROTECTION FOR INDIVIDUAL OVENS.

234

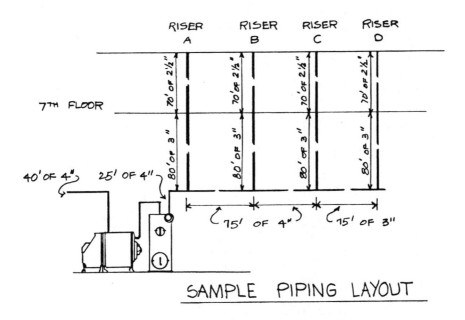

RISER A    RISER B    RISER C    RISER D

70' OF 2½"   70' OF 2½"   70' OF 2½"   70' OF 2½"

7TH FLOOR

80' OF 3"   80' OF 3"   80' OF 3"   80' OF 3"

40' OF 4"    25' OF 4"

75' OF 4"    75' OF 3"

## SAMPLE PIPING LAYOUT

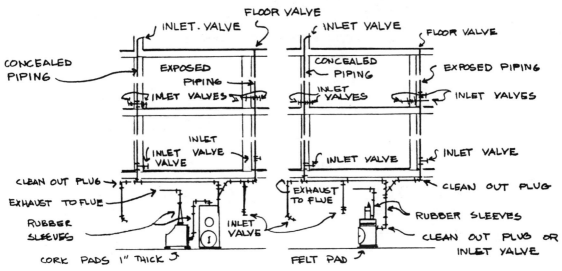

FLOOR VALVE

INLET. VALVE    INLET VALVE    FLOOR VALVE

CONCEALED PIPING    EXPOSED PIPING    CONCEALED PIPING    EXPOSED PIPING

INLET VALVES    INLET VALVES    INLET VALVES

INLET INLET VALVE VALVE    INLET VALVE    INLET VALVE

CLEAN OUT PLUG    CLEAN OUT PLUG

EXHAUST TO FLUE    EXHAUST TO FLUE    RUBBER SLEEVES

RUBBER SLEEVES    INLET VALVE    CLEAN OUT PLUG OR INLET VALVE

CORK PADS 1" THICK    FELT PAD

## FIGURE 2-55

## CENTRAL VACUUM SYSTEM

Figures 2-55 and 2-56 cover two schematic arrangements of seldom used details. It has been the author's experience in practice over many years that central vacuum systems are not used very much in consultant practice. However, there are times when this type of system is required and these two details are included for general information.

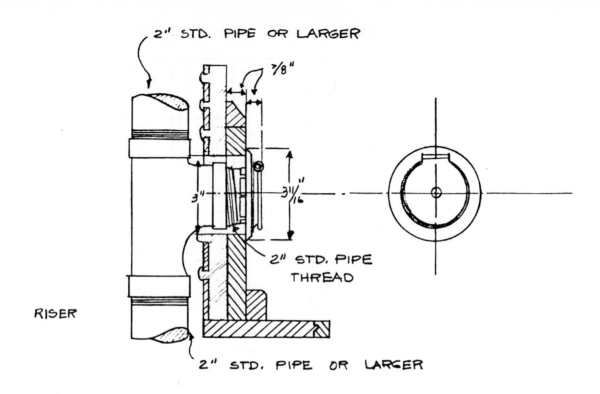

2" STD. PIPE OR LARGER

⅞"

3"

3¹¹⁄₁₆"

2" STD. PIPE THREAD

RISER

2" STD. PIPE OR LARGER

TYPICAL INLET VALVE

**FIGURE 2-56**

CENTRAL VACUUM SYSTEM

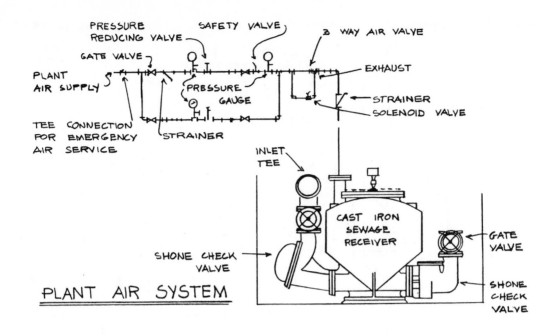

PLANT AIR SYSTEM

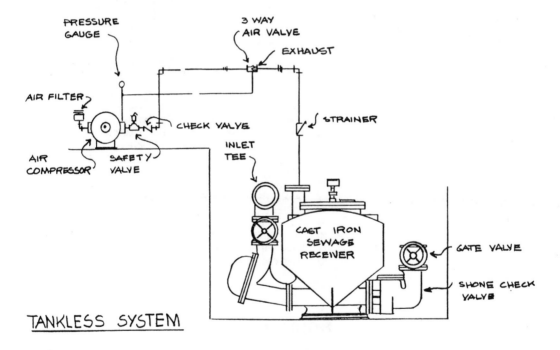

TANKLESS SYSTEM

**FIGURE 2-57**

PNEUMATIC EJECTOR

Figure 2-57 illustrates a little used system that easily is forgotten, omitted, or not covered. That system is air piping to a sewage receiver, using either a plant air system or an air compressor. The detail is self-explanatory.

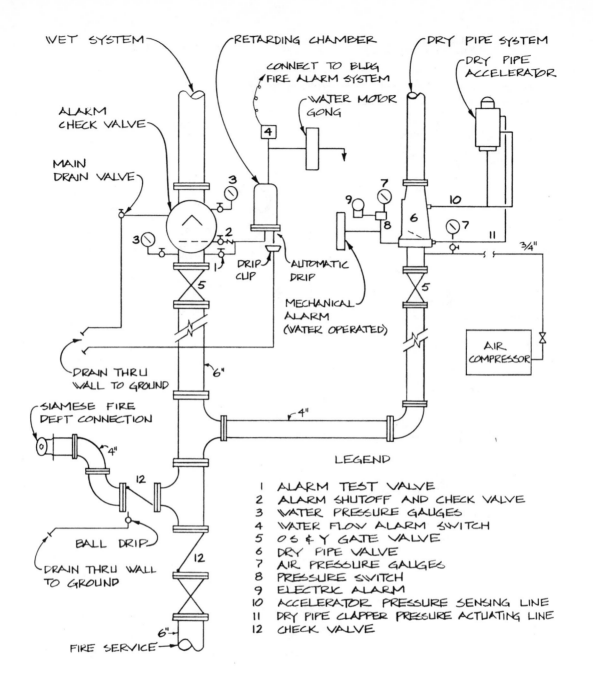

**FIGURE 2-58**

SPRINKLER SERVICE

NO SCALE

WET SYSTEM

RETARDING CHAMBER

CONNECT TO BLDG FIRE ALARM SYSTEM

DRY PIPE SYSTEM

DRY PIPE ACCELERATOR

ALARM CHECK VALVE

WATER MOTOR GONG

MAIN DRAIN VALVE

DRIP CUP

AUTOMATIC DRIP

MECHANICAL ALARM (WATER OPERATED)

DRAIN THRU WALL TO GROUND

6"

4"

AIR COMPRESSOR

3/4"

SIAMESE FIRE DEPT CONNECTION

4"

12

BALL DRIP

DRAIN THRU WALL TO GROUND

12

6"

FIRE SERVICE

LEGEND

1  ALARM TEST VALVE
2  ALARM SHUTOFF AND CHECK VALVE
3  WATER PRESSURE GAUGES
4  WATER FLOW ALARM SWITCH
5  O S & Y GATE VALVE
6  DRY PIPE VALVE
7  AIR PRESSURE GAUGES
8  PRESSURE SWITCH
9  ELECTRIC ALARM
10 ACCELERATOR PRESSURE SENSING LINE
11 DRY PIPE CLAPPER PRESSURE ACTUATING LINE
12 CHECK VALVE

Figure 2-58 is a composite drawing of a sprinkler service having both a dry pipe and a wet pipe requirement. The sizes, of course, would vary with the type of project and the number of heads. The dry pipe could or could not be eliminated. The wet pipe could or could not be eliminated. The detail can be used in part or as is, depending on the application requirements.

**238**

| PLUMBING FIXTURE SCHEDULE | | | | | | | |
|---|---|---|---|---|---|---|---|
| FIXTURE | SIZE - TYPE | MANUFACTURER AND MODEL | SOIL | VENT | CW | HW | REMARKS, ACCESSORIES, NOTES |
| | | | | | | | |
| | | | | | | | |
| | | | | | | | |
| | | | | | | | |

| PLUMBING EQUIPMENT SCHEDULE | | | | | | | | | | |
|---|---|---|---|---|---|---|---|---|---|---|
| ITEM | SIZE - TYPE | MANUFACTURER AND MODEL | SOIL | VENT | CW | HW | RATING | INPUT | OUTPUT | NOTES |
| | | | | | | | | | | |
| | | | | | | | | | | |
| | | | | | | | | | | |
| | | | | | | | | | | |

## FIGURE 2-59

MECHANICAL SCHEDULES

Figure 2-59 consists of schedules, which we have found to be the only ones of much practical use for plumbing work. The plumbing fixtures schedule is used constantly since the fixture list is not put in the specifications. The equipment schedule varies from job to job—sometimes it is used, sometimes not. Its use is really up to individual discretion.

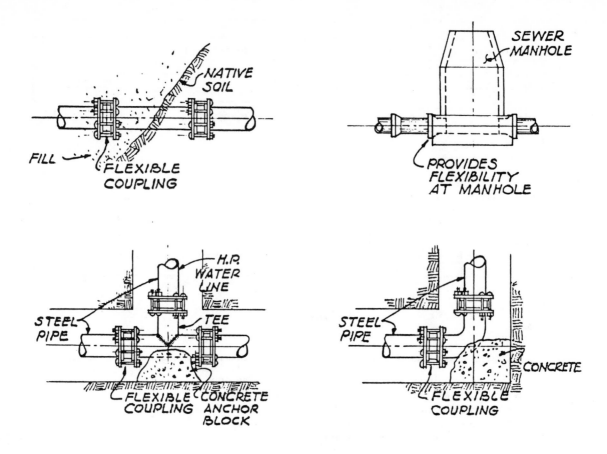

**FIGURE 2-60**

TYPICAL SEISMIC PIPING CONNECTIONS

1 OF 3

Figures 2-60, 2-61, and 2-62 are typical piping connections which provide flexibility for seismic conditions. None of these details should be construed as eliminating the need for a thorough analysis. To prove that the piping system will withstand the ground acceleration that may occur in a major seismic event, the concrete must not interfere with the coupling. Anchor tees require three flexible couplings. Pipes crossing a line between fill and natural soil require flexible couplings, as shown. Flexibility should be provided as close to a manhole as possible. Flexible connections are not necessary where anchor blocks are not used. Finally, there are various arrangements shown for pipe flexibility on Figure 2-62.

240

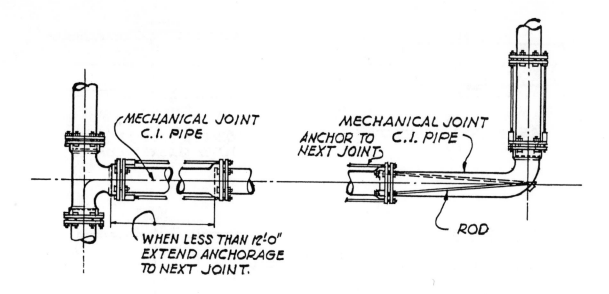

MECHANICAL JOINT C.I. PIPE

WHEN LESS THAN 12'-0" EXTEND ANCHORAGE TO NEXT JOINT.

MECHANICAL JOINT C.I. PIPE

ANCHOR TO NEXT JOINT

ROD

_TEE_

_90° BEND_

**FIGURE 2-61**

_TYPICAL SEISMIC PIPING CONNECTIONS_
_2 OF 3_

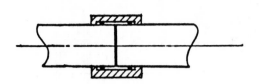

ASBESTOS·CEMENT
COUPLING

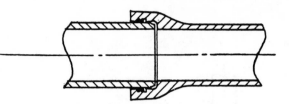

BELL & SPIGOT JOINT
WITH GASKET CONNECTION

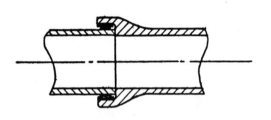

VCP CONNECTION

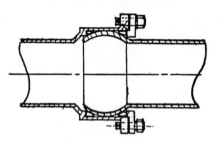

BALL JOINT

FIGURE 2-62

TYPICAL SEISMIC PIPING CONNECTIONS
3 OF 3

242

# 3. Electrical Systems

**244**

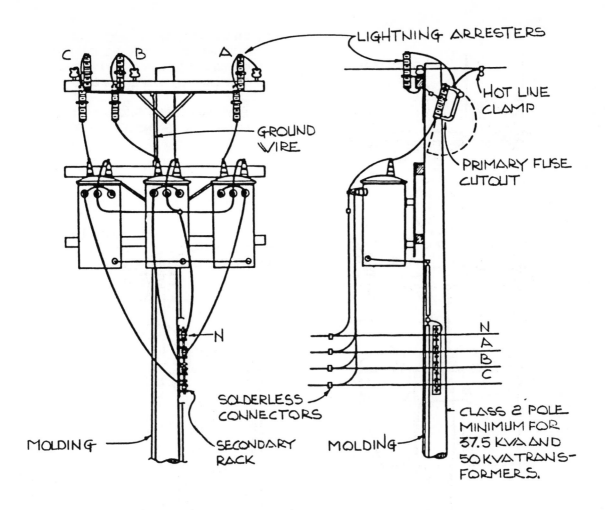

**FIGURE 3-1**

TYPICAL THREE PHASE TRANSFORMER BANK
CROSSARM MOUNTING
FOR 3-50 KVA OR LESS

Figures 3-1 and 3-2 are typical light duty transformer arrangements on a pole. The first drawing shows the crossarm arrangement and the second the cluster arrangement. Note that the transformers are mounted between the primary and the secondary with short leads on both sides of the transformer. The proper sizes of the various items must be carefully calculated and the pole must be an approved type. Generally, 20% of the pole height above the ground is the depth of pole in the ground.

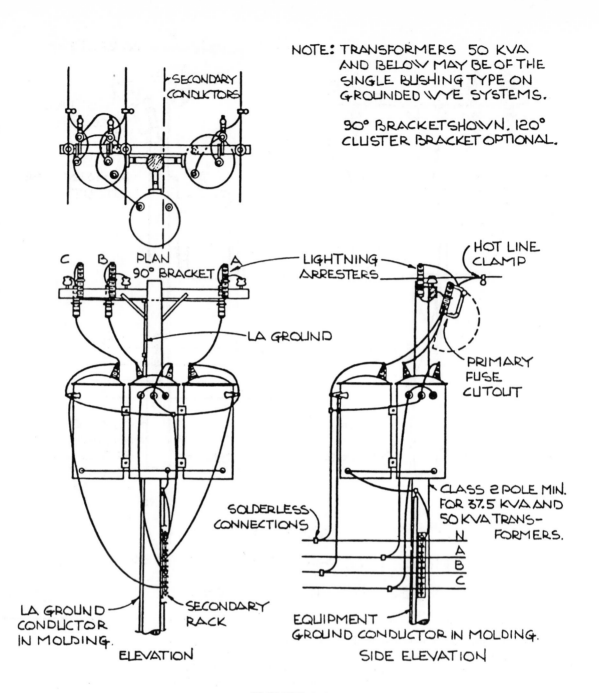

NOTE: TRANSFORMERS 50 KVA
AND BELOW MAY BE OF THE
SINGLE BUSHING TYPE ON
GROUNDED WYE SYSTEMS.

90° BRACKET SHOWN. 120°
CLUSTER BRACKET OPTIONAL.

SECONDARY
CONDUCTORS

C    B    PLAN         A
     90° BRACKET

LIGHTNING
ARRESTERS

HOT LINE
CLAMP

LA GROUND

PRIMARY
FUSE
CUTOUT

SOLDERLESS
CONNECTIONS

CLASS 2 POLE MIN.
FOR 37.5 KVA AND
50 KVA TRANS-
FORMERS.

N
A
B
C

LA GROUND
CONDUCTOR
IN MOLDING.

SECONDARY
RACK

EQUIPMENT
GROUND CONDUCTOR IN MOLDING.

ELEVATION

SIDE ELEVATION

**FIGURE 3-2**

TYPICAL THREE PHASE TRANSFORMER BANK
CLUSTER MOUNT
3-50 KVA OR LESS

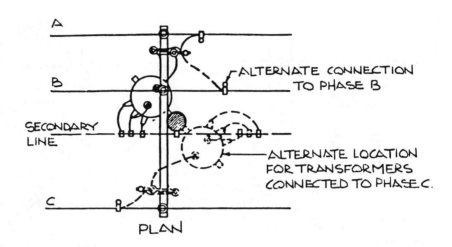

A

B

ALTERNATE CONNECTION
TO PHASE B

SECONDARY
LINE

ALTERNATE LOCATION
FOR TRANSFORMERS
CONNECTED TO PHASE C.

C

PLAN

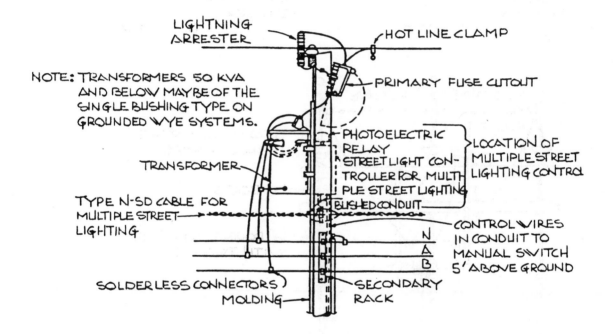

LIGHTNING
ARRESTER

HOT LINE CLAMP

NOTE: TRANSFORMERS 50 KVA
AND BELOW MAY BE OF THE
SINGLE BUSHING TYPE ON
GROUNDED WYE SYSTEMS.

PRIMARY FUSE CUTOUT

PHOTO ELECTRIC
RELAY
STREET LIGHT CON-
TROLLER FOR MULTI-
PLE STREET LIGHTING

LOCATION OF
MULTIPLE STREET
LIGHTING CONTROL

TRANSFORMER

TYPE N-SD CABLE FOR
MULTIPLE STREET
LIGHTING

BUSHED CONDUIT

CONTROL WIRES
IN CONDUIT TO
MANUAL SWITCH
5' ABOVE GROUND

N
A
B

SOLDERLESS CONNECTORS
MOLDING

SECONDARY
RACK

### FIGURE 3-3

TYPICAL SINGLE-PHASE TRANSFORMER INSTALLATION

Figure 3-3 is similar to Figures 3-1 and 3-2, except it is for a single transformer installation and is shown with a street lighting controller, which
may or may not be a part of the work.

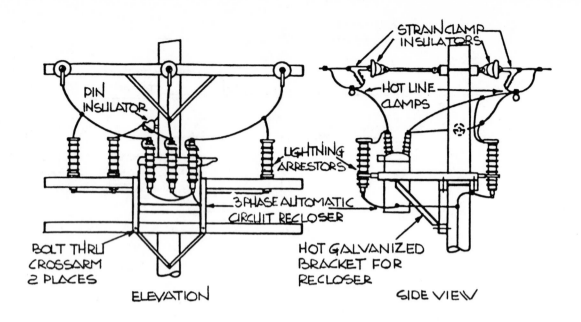

PIN INSULATOR

STRAIN CLAMP INSULATORS

HOT LINE CLAMPS

LIGHTNING ARRESTORS

3 PHASE AUTOMATIC CIRCUIT RECLOSER

BOLT THRU CROSSARM 2 PLACES

HOT GALVANIZED BRACKET FOR RECLOSER

ELEVATION

SIDE VIEW

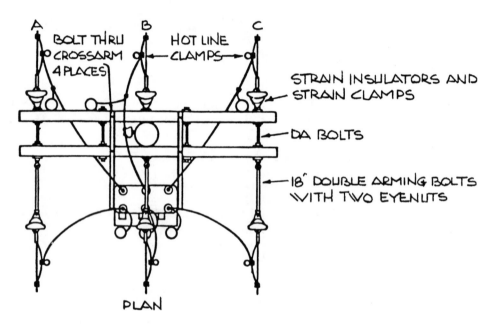

A   B   C

BOLT THRU CROSSARM 4 PLACES

HOT LINE CLAMPS

STRAIN INSULATORS AND STRAIN CLAMPS

DA BOLTS

18" DOUBLE ARMING BOLTS WITH TWO EYENUTS

PLAN

**FIGURE 3-4**

TYPICAL INSTALLATION
3-PHASE AUTOMATIC CIRCUIT RECLOSER

---

Figure 3-4 illustrates a typical primary type automatic circuit recloser. The purpose of the recloser is to restore service quickly after temporary faults on overhead open wire circuits have been cleared. Most temporary faults will clear themselves after the first opening of the primary breaker. The first reclosure is made without any time delay to restore service as quickly as possible. Exact specifications on this system are up to the designer involved.

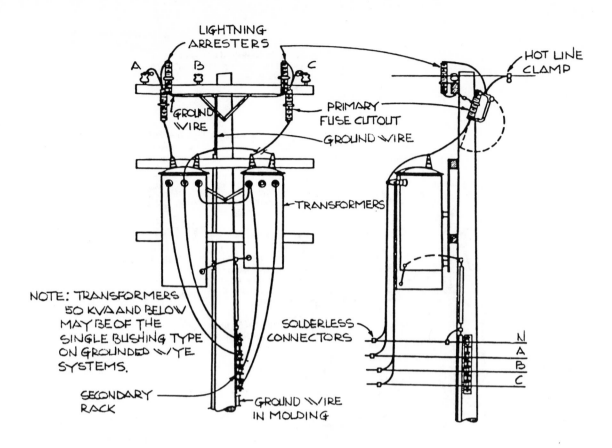

**FIGURE 3-5**

TYPICAL THREE - PHASE TRANSFORMER
INSTALLATION. OPEN DELTA WITH TAP
AT MIDPOINT OF ONE TRANSFORMER
CROSS ARM MOUNT.

Figure 3-5 is also similar to Figures 3-1 and 3-2, except that it shows the relationship of an open delta type arrangement with two transformers.

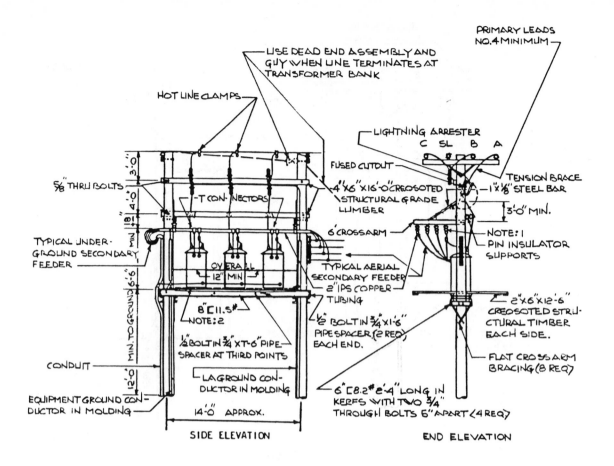

USE DEAD END ASSEMBLY AND
GUY WHEN LINE TERMINATES AT
TRANSFORMER BANK

HOT LINE CLAMPS

PRIMARY LEADS
NO. 4 MINIMUM

LIGHTNING ARRESTER
C  SL  B  A

FUSED CUTOUT

5/8" THRU BOLTS

T CON NECTORS

4"X6"X16'-0" CREOSOTED
STRUCTURAL GRADE
LUMBER

TENSION BRACE
1"X 1/8" STEEL BAR

3'-0" MIN.

6' CROSSARM

NOTE: 1
PIN INSULATOR
SUPPORTS

TYPICAL UNDER-
GROUND SECONDARY
FEEDER

OVERALL
12" MIN

TYPICAL AERIAL
SECONDARY FEEDER
2" IPS COPPER
TUBING

8"C11.5#
NOTE: 2

2"X6"X12'-6"
CREOSOTED STRU-
CTURAL TIMBER
EACH SIDE.

1/2" BOLT IN 3/4"X1'-6"
PIPE SPACER (2 REQ)
EACH END.

1/2" BOLT IN 3/4"XT'-6" PIPE-
SPACER AT THIRD POINTS

FLAT CROSSARM
BRACING (8 REQ)

CONDUIT

LA GROUND CON-
DUCTOR IN MOLDING

EQUIPMENT GROUND CON-
DUCTOR IN MOLDING

14'-0" APPROX.

6"C8.2# 2'-4" LONG IN
KERFS WITH TWO 3/4"
THROUGH BOLTS 5" APART (4 REQ)

SIDE ELEVATION

END ELEVATION

NOTES: 1. INSULATOR BUS SUPPORTS SHALL BE NEMA TYPE TR-1 APPARATUS CAP
AND PIN INSULATOR WITH TOP CLAMP FOR COPPER TUBBING BUS. BOLT
TO CROSSARM WITH TWO 1/2" THROUGH BOLTS.

2. SIXTEEN 2"X8"X8'-0" CREOSOTED STRUCTURAL GRADE TIMBERS ON 10±
CENTERS HELD WITH A 5/8" X 3 1/2" CARRIAGE BOLT IN EACH CHANNEL.

3. TRANSFORMERS OR TRANSFORMER BANK HAVING A CAPACITY IN EXCESS
OF 501 KVA SHALL BE MOUNTED ON A FENCE ENCLOSED CONCRETE MAT.
CONNECTIONS TO BE DELTA OR WYE AS REQUIRED. PRIMARY AND
SECONDARY BUS NEUTRALS MAY BE COMBINED.

## FIGURE 3-6

TYPICAL THREE PHASE TRANSFORMER BANK

3-75 KVA TO 3-167 KVA, INCL.

Figure 3-6 is another version of larger transformers that are rack mounted.
The notes on the detail are self-explanatory. Attention is called to the 500-
kVA limitation.

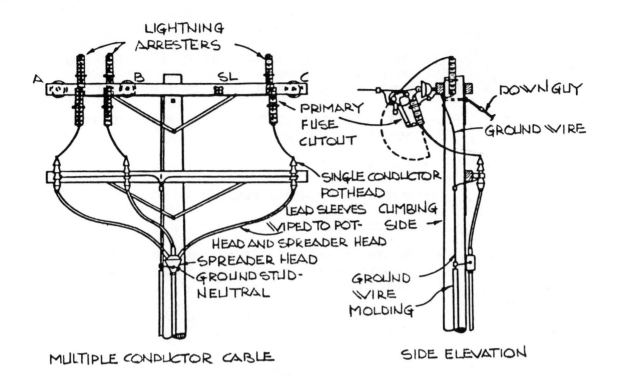

LIGHTNING ARRESTERS

A    B    SL    C

PRIMARY FUSE CUTOUT

SINGLE CONDUCTOR POTHEAD

LEAD SLEEVES WIPED TO POTHEAD AND SPREADER HEAD

SPREADER HEAD

GROUND STUD-NEUTRAL

MULTIPLE CONDUCTOR CABLE

DOWN GUY

GROUND WIRE

CLIMBING SIDE

GROUND WIRE MOLDING

SIDE ELEVATION

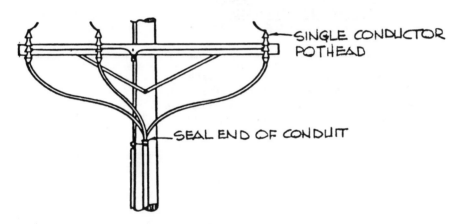

SINGLE CONDUCTOR POTHEAD

SEAL END OF CONDUIT

SINGLE CONDUCTOR CABLE

## FIGURE 3-7

### TYPICAL CONNECTION—AERIAL TO UNDERGROUND

Figure 3-7 is a detail of the connection of aerial to underground cable and it illustrates ways of connecting single and multiple cables.

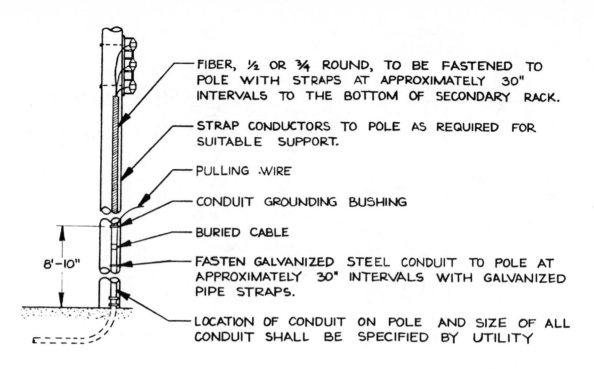

FIBER, ½ OR ¾ ROUND, TO BE FASTENED TO POLE WITH STRAPS AT APPROXIMATELY 30" INTERVALS TO THE BOTTOM OF SECONDARY RACK.

STRAP CONDUCTORS TO POLE AS REQUIRED FOR SUITABLE SUPPORT.

PULLING WIRE

CONDUIT GROUNDING BUSHING

BURIED CABLE

FASTEN GALVANIZED STEEL CONDUIT TO POLE AT APPROXIMATELY 30" INTERVALS WITH GALVANIZED PIPE STRAPS.

LOCATION OF CONDUIT ON POLE AND SIZE OF ALL CONDUIT SHALL BE SPECIFIED BY UTILITY

8'-10"

## FIGURE 3-8

### POLE RISER INSTALLATION
### SECONDARY SERVICE
### NO SCALE

---

Figure 3-8 is, in effect, a continuation of Figure 3-7 in that it shows the idea of running aerial to underground cables, only in this case the underground portion is shown. Approximate dimension of the conduit at the pole is noted on the plans. Again, the sizing is up to the engineer.

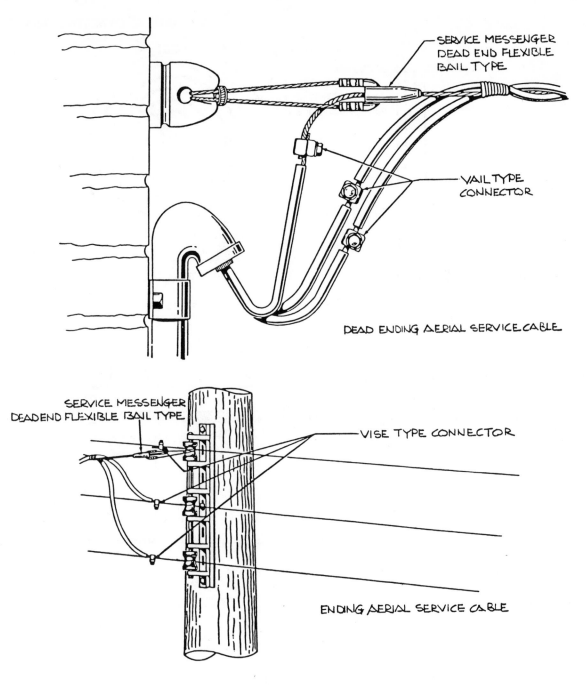

SERVICE MESSENGER DEAD END FLEXIBLE BAIL TYPE

VAIL TYPE CONNECTOR

DEAD ENDING AERIAL SERVICE CABLE

SERVICE MESSENGER DEADEND FLEXIBLE BAIL TYPE

VISE TYPE CONNECTOR

ENDING AERIAL SERVICE CABLE

**FIGURE 3-9**

ENDING AERIAL SERVICE CABLE

Figure 3-9 is relatively simple. It is merely an enlarged detail of ending an aerial service cable, which is sometimes used on various final drawings. Materials may be noted on the detail.

255

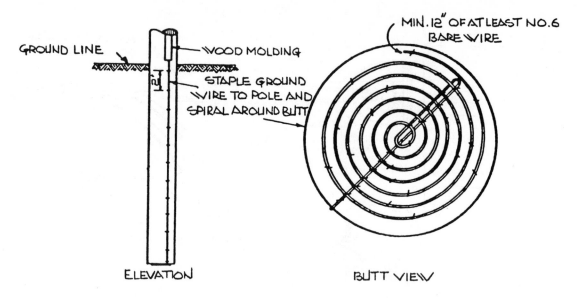

GROUND LINE

WOOD MOLDING

STAPLE GROUND
WIRE TO POLE AND
SPIRAL AROUND BUTT

2'

ELEVATION

MIN. 12" OF AT LEAST NO. 6
BARE WIRE

BUTT VIEW

TYPICAL COIL-TYPE GROUND (FOR USE WHERE GROUND RESISTANCE IS LOW)

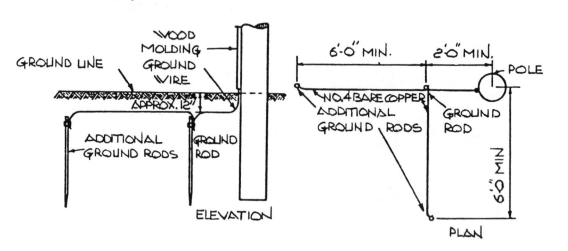

GROUND LINE

WOOD MOLDING GROUND WIRE

APPROX. 12"

ADDITIONAL GROUND RODS

GROUND ROD

ELEVATION

6'-0" MIN.

2'-0" MIN.

POLE

NO. 4 BARE COPPER
ADDITIONAL GROUND RODS

GROUND ROD

6'0" MIN.

PLAN

**FIGURE 3-10**

Figure 3-10 illustrates some common ways of grounding for a typical ground rod. Obviously, the actual ground requirements would have to be calculated, but the detail shown is generally standard.

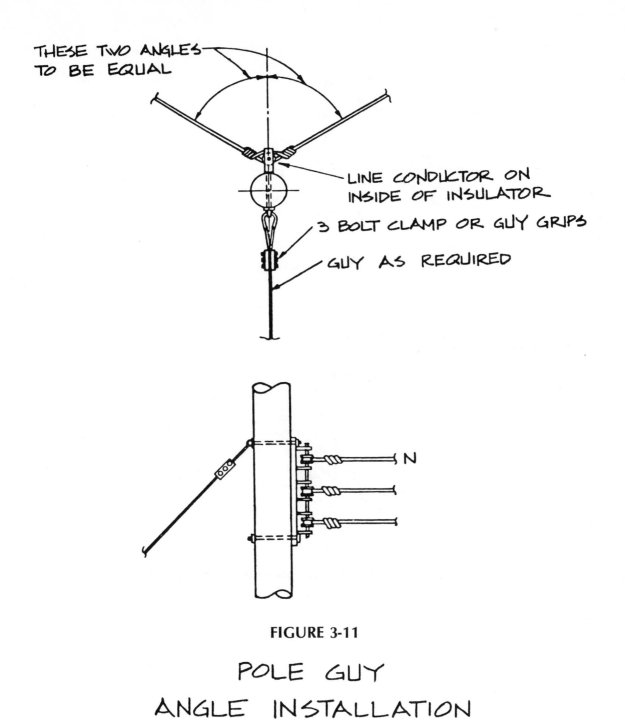

THESE TWO ANGLES
TO BE EQUAL

LINE CONDUCTOR ON
INSIDE OF INSULATOR

3 BOLT CLAMP OR GUY GRIPS

GUY AS REQUIRED

N

FIGURE 3-11

POLE GUY
ANGLE INSTALLATION

Figures 3-11 through 3-14 are typical details of overhead wiring which,
from time to time, must be shown on the drawings. In all cases, the type
of wire, size of wire, and materials to be used in the various connecting
devices would have to be specified.

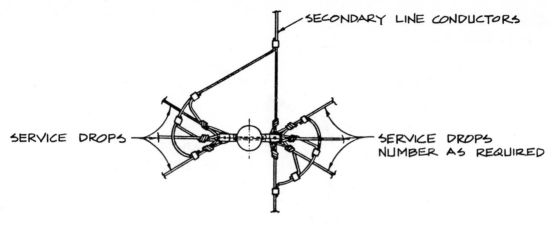

SECONDARY LINE CONDUCTORS

SERVICE DROPS

SERVICE DROPS
NUMBER AS REQUIRED

OPEN WIRE SERVICE

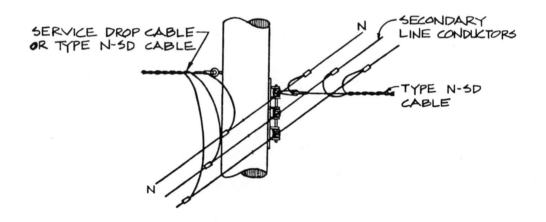

SERVICE DROP CABLE
OR TYPE N-SD CABLE

SECONDARY
LINE CONDUCTORS

TYPE N-SD
CABLE

**FIGURE 3-12**

TYPICAL SERVICE CONNECTIONS

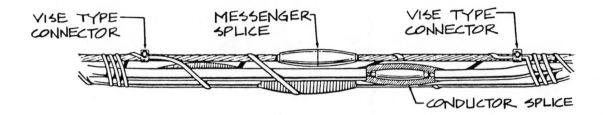

VISE TYPE CONNECTOR    MESSENGER SPLICE    VISE TYPE CONNECTOR

CONDUCTOR SPLICE

METHOD OF SPLICING AERIAL CABLE
MESSENGER & CONDUCTORS

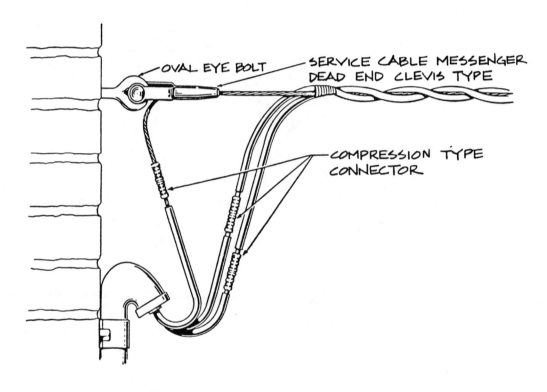

OVAL EYE BOLT    SERVICE CABLE MESSENGER
DEAD END CLEVIS TYPE

COMPRESSION TYPE
CONNECTOR

DEAD ENDING AERIAL SERVICE CABLE

**FIGURE 3-13**

AERIAL CABLE SPLICE & SERVICE ENTRY

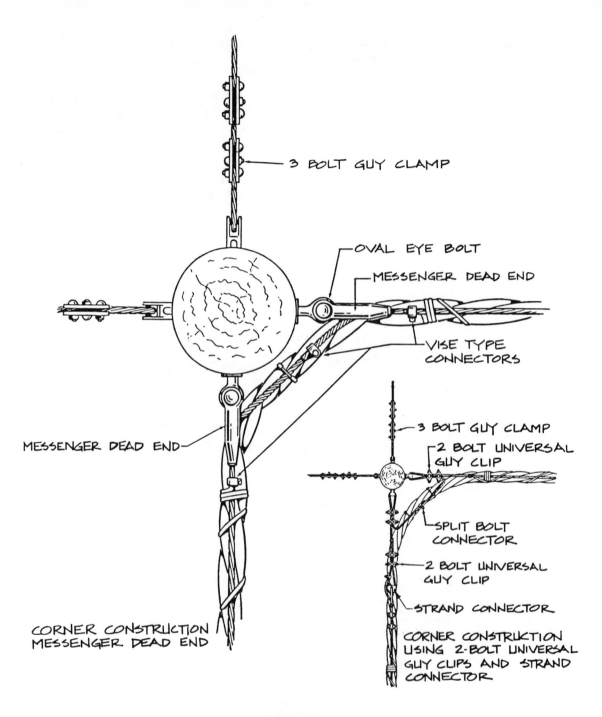

3 BOLT GUY CLAMP

OVAL EYE BOLT

MESSENGER DEAD END

VISE TYPE CONNECTORS

MESSENGER DEAD END

CORNER CONSTRUCTION MESSENGER DEAD END

3 BOLT GUY CLAMP

2 BOLT UNIVERSAL GUY CLIP

SPLIT BOLT CONNECTOR

2 BOLT UNIVERSAL GUY CLIP

STRAND CONNECTOR

CORNER CONSTRUCTION USING 2-BOLT UNIVERSAL GUY CLIPS AND STRAND CONNECTOR

**FIGURE 3-14**

AERIAL CABLE CORNER CONSTRUCTION

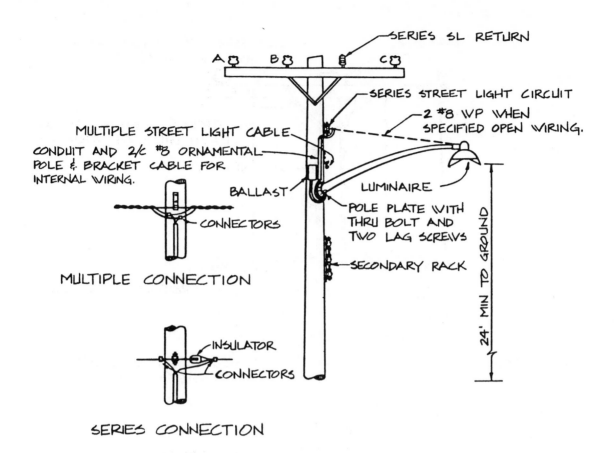

SERIES SL RETURN

A □ B □ C □

SERIES STREET LIGHT CIRCUIT

2 #8 WP WHEN SPECIFIED OPEN WIRING.

MULTIPLE STREET LIGHT CABLE

CONDUIT AND 2/c #8 ORNAMENTAL POLE & BRACKET CABLE FOR INTERNAL WIRING.

BALLAST

LUMINAIRE

POLE PLATE WITH THRU BOLT AND TWO LAG SCREWS

CONNECTORS

SECONDARY RACK

24' MIN TO GROUND

MULTIPLE CONNECTION

INSULATOR

CONNECTORS

SERIES CONNECTION

**FIGURE 3-15**

TYPICAL STREET LIGHTING UNIT

NOT TO SCALE

---

Figures 3-15 and 3-16 are, respectively, typical street lighting units and typical street lighting regulators. Figure 3-15 illustrates the street lighting unit with some of the component wiring not shown for greater clarity. Figure 3-16 shows the regulator and wiring. All items require sizing by the engineer.

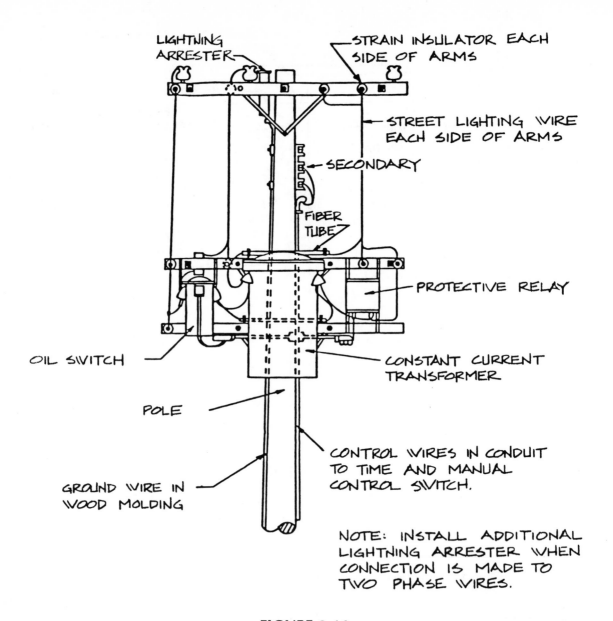

LIGHTNING ARRESTER

STRAIN INSULATOR EACH SIDE OF ARMS

STREET LIGHTING WIRE EACH SIDE OF ARMS

SECONDARY

FIBER TUBE

PROTECTIVE RELAY

OIL SWITCH

CONSTANT CURRENT TRANSFORMER

POLE

CONTROL WIRES IN CONDUIT TO TIME AND MANUAL CONTROL SWITCH.

GROUND WIRE IN WOOD MOLDING

NOTE: INSTALL ADDITIONAL LIGHTNING ARRESTER WHEN CONNECTION IS MADE TO TWO PHASE WIRES.

**FIGURE 3-16**

STREET LIGHTING REGULATOR INSTALLATION

NOT TO SCALE

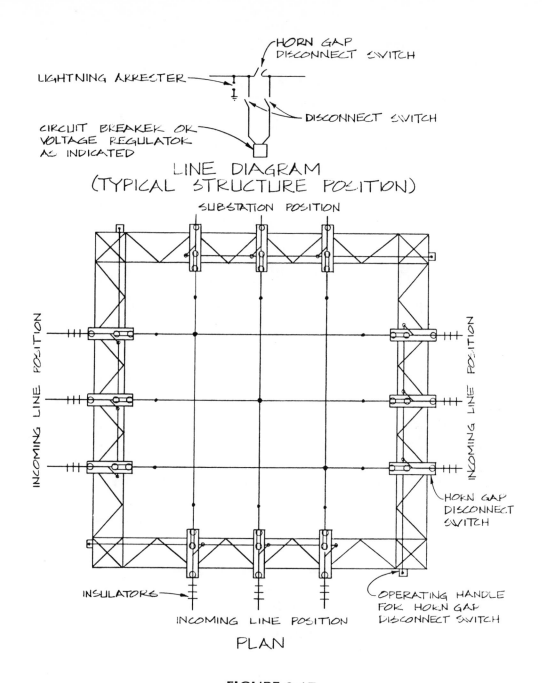

HORN GAP
DISCONNECT SWITCH

LIGHTNING ARRESTER

DISCONNECT SWITCH

CIRCUIT BREAKER OR
VOLTAGE REGULATOR
AS INDICATED

LINE DIAGRAM
(TYPICAL STRUCTURE POSITION)

SUBSTATION POSITION

INCOMING LINE POSITION

INCOMING LINE POSITION

HORN GAP
DISCONNECT
SWITCH

INSULATORS

INCOMING LINE POSITION

OPERATING HANDLE
FOR HORN GAP
DISCONNECT SWITCH

PLAN

**FIGURE 3-17**

TYPICAL MULTI-CIRCUIT
PRIMARY OR SECONDARY STRUCTURE
NO SCALE

Figures 3-17 through 3-21 primarily show details of the structure involved
in a high voltage main distribution outdoor installation. These structures
would, in each case, have to be carefully calculated for member sizing
foundations and the like by the structural engineer. They can be used in
offices as guides to be further detailed.

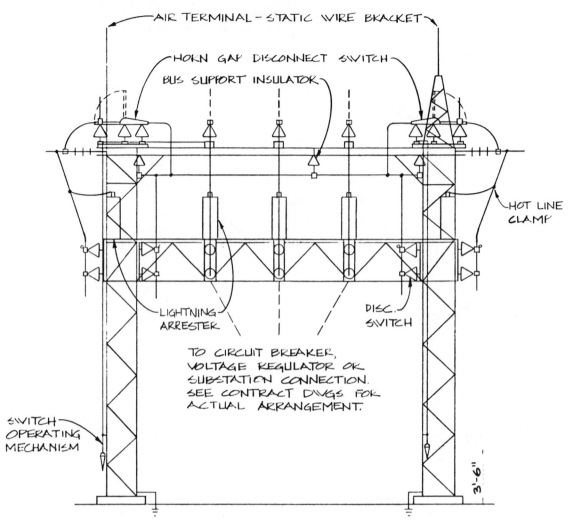

ELEVATION

**FIGURE 3-18**

TYPICAL MULTI-CIRCUIT
PRIMARY OR SECONDARY STRUCTURE

NO SCALE

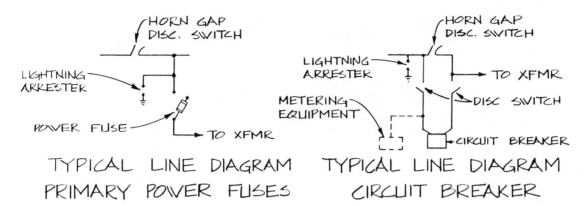

HORN GAP
DISC. SWITCH

LIGHTNING
ARRESTER

POWER FUSE

TO XFMR

TYPICAL LINE DIAGRAM
PRIMARY POWER FUSES

HORN GAP
DISC. SWITCH

LIGHTNING
ARRESTER

TO XFMR

METERING
EQUIPMENT

DISC SWITCH

CIRCUIT BREAKER

TYPICAL LINE DIAGRAM
CIRCUIT BREAKER

HORN GAP
DISCONNECT SWITCH

INSULATOR

POWER
FUSE

LIGHTNING
ARRESTER

TO TRANSFORMER

POWER FUSE DISC. SWITCH

HORN GAP
DISCONNECT SWITCH

HOT LINE
CLAMP

TO XFMR

DISC. SWITCH

CIRCUIT BREAKER

**FIGURE 3-19**

TYPICAL SINGLE CIRCUIT - PRIMARY OR SECONDARY
STRUCTURE - SIDE ELEVATIONS
NO SCALE

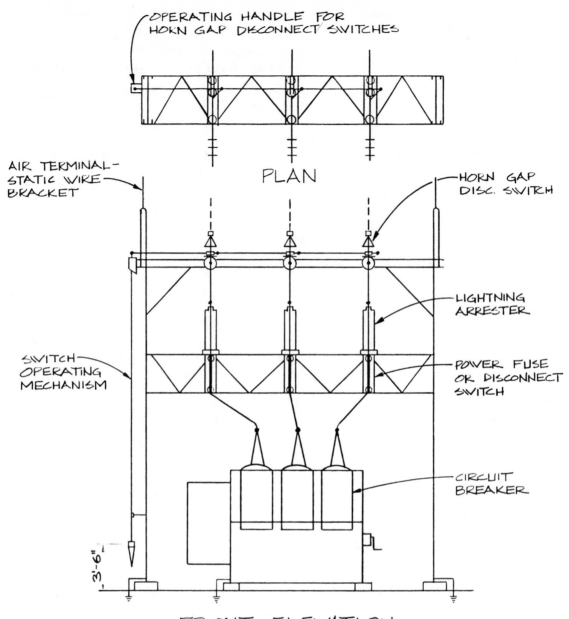

OPERATING HANDLE FOR HORN GAP DISCONNECT SWITCHES

PLAN

AIR TERMINAL-STATIC WIRE BRACKET

HORN GAP DISC. SWITCH

SWITCH OPERATING MECHANISM

LIGHTNING ARRESTER

POWER FUSE OR DISCONNECT SWITCH

CIRCUIT BREAKER

3'-6"

FRONT ELEVATION

**FIGURE 3-20**

TYPICAL SINGLE CIRCUIT
PRIMARY OR SECONDARY STRUCTURE

NO SCALE

266

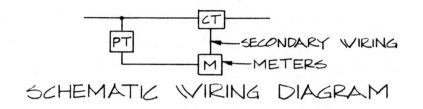

SCHEMATIC WIRING DIAGRAM

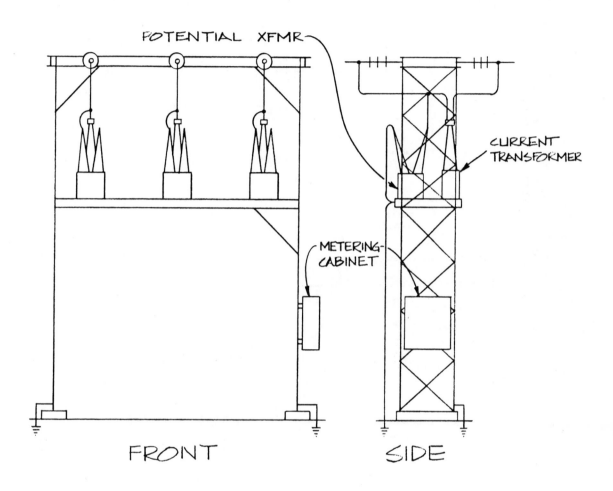

FRONT          SIDE

**FIGURE 3-21**

TYPICAL PRIMARY METERING STRUCTURE

NO SCALE

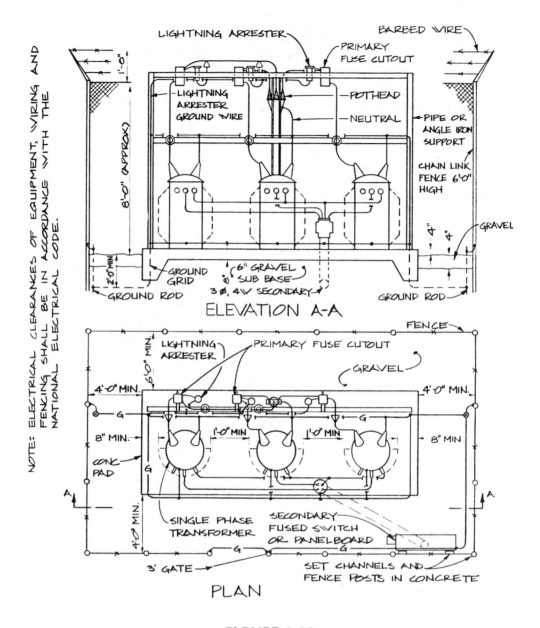

NOTE: ELECTRICAL CLEARANCES OF EQUIPMENT, WIRING AND FENCING SHALL BE IN ACCORDANCE WITH THE NATIONAL ELECTRICAL CODE.

ELEVATION A-A

PLAN

**FIGURE 3-22**

TYPICAL THREE PHASE OUTDOOR
CONVENTIONAL TRANSFORMER INSTALLATION
NO SCALE

Figures 3-22 and 3-23 are typical outdoor three phase transformer installations. There are other details in this section, which show the pad and some of the grounding and wiring associated with it. In this case, the work above the pad, the general arrangement of potheads, overhead wiring, and transformer primary and secondary wiring are shown. In each case, all of the members would have to be sized, and the primary and secondary wiring would also have to be sized. The clearances from the pad to the fence are important to note.

268

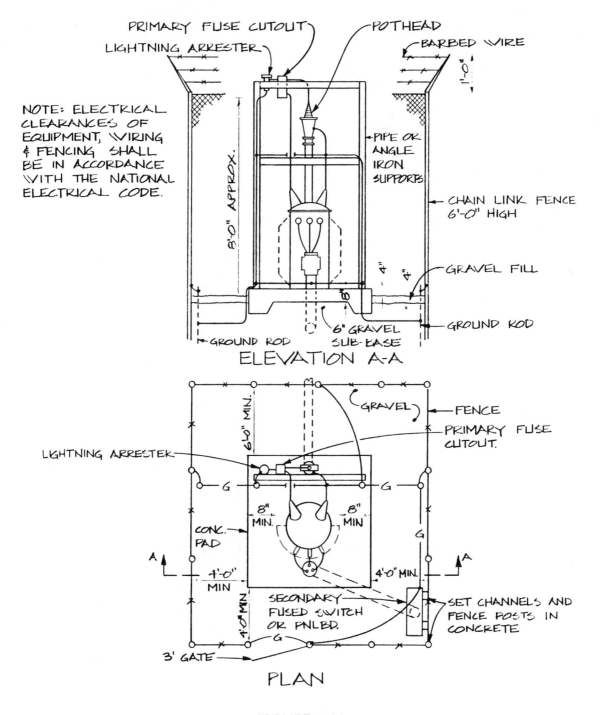

PRIMARY FUSE CUTOUT
LIGHTNING ARRESTER
POTHEAD
BARBED WIRE

NOTE: ELECTRICAL CLEARANCES OF EQUIPMENT, WIRING & FENCING SHALL BE IN ACCORDANCE WITH THE NATIONAL ELECTRICAL CODE.

8'-0" APPROX.

1'-0"

PIPE OR ANGLE IRON SUPPORTS

CHAIN LINK FENCE 6'-0" HIGH

GRAVEL FILL

4"  4"

GROUND ROD

GROUND ROD          6" GRAVEL SUB-BASE

ELEVATION A-A

GRAVEL
FENCE
PRIMARY FUSE CUTOUT.

6'-0" MIN.

LIGHTNING ARRESTER

G          G

G

8" MIN.          8" MIN.

CONC. PAD

A

4'-0" MIN

4'-0" MIN.

SECONDARY FUSED SWITCH OR PNLBD.

SET CHANNELS AND FENCE POSTS IN CONCRETE

A

4'-0" MIN

G

3' GATE

PLAN

**FIGURE 3-23**

TYPICAL SINGLE PHASE OUTDOOR CONVENTIONAL TRANSFORMER INSTALLATION
NO SCALE

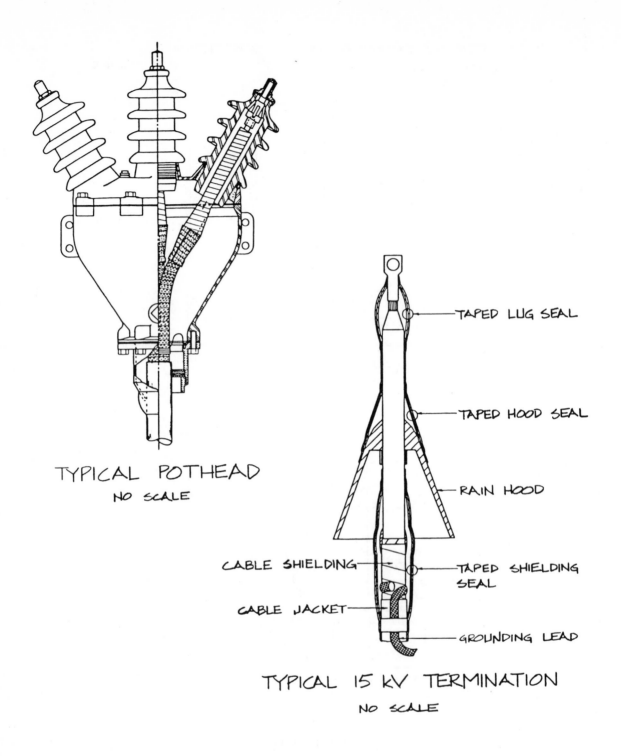

TYPICAL POTHEAD

NO SCALE

TAPED LUG SEAL

TAPED HOOD SEAL

RAIN HOOD

CABLE SHIELDING

TAPED SHIELDING SEAL

CABLE JACKET

GROUNDING LEAD

TYPICAL 15 kV TERMINATION

NO SCALE

**FIGURE 3-24**

Figures 3-24 and 3-25 are details of 15,000-V termination and connections that are frequently required when doing 15-kV work. These details are enlarged and show a typical pothead, typical termination, and a typical pipe splice. Obviously, these details would vary somewhat with different types of wire.

270

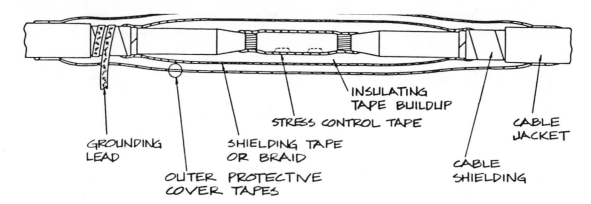

INSULATING
TAPE BUILDUP

STRESS CONTROL TAPE

CABLE
JACKET

GROUNDING
LEAD

SHIELDING TAPE
OR BRAID

CABLE
SHIELDING

OUTER PROTECTIVE
COVER TAPES

NOTE: HEAVY BRAID JUMPER SHOULD BE USED ACROSS SPLICE
TO CARRY POSSIBLE GROUND-FAULT CURRENT. STRESS
CONTROL TAPE SHOULD COVER STRANDS COMPLETELY,
LAPPING SLIGHTLY ONTO INSULATION TAPER.

## TYPICAL TAPED SPLICE IN SHIELDED CABLE

NO SCALE

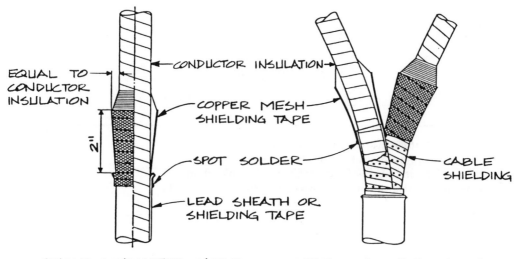

EQUAL TO
CONDUCTOR
INSULATION

CONDUCTOR INSULATION

COPPER MESH
SHIELDING TAPE

SPOT SOLDER

CABLE
SHIELDING

2"

LEAD SHEATH OR
SHIELDING TAPE

SINGLE CONDUCTOR CABLE

MULTI-CONDUCTOR CABLE

## TYPICAL STRESS RELIEF CONES

NO SCALE

**FIGURE 3-25**

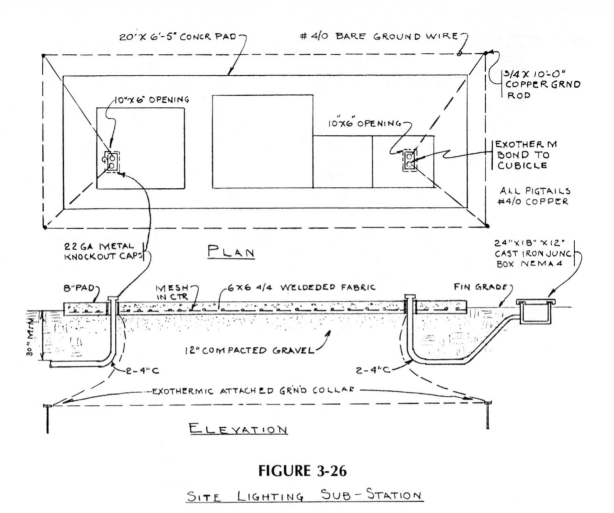

**FIGURE 3-26**

SITE LIGHTING SUB-STATION

Figure 3-26 is a detail for the construction of the pad and the underground work in connection with a substation. In this particular case, it is taken from an actual job, and obviously all of the sizes are based on that job. For the particular job that the engineer may have, these sizes would, of course, vary.

272

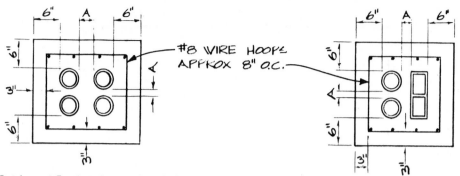

TYPICAL SECTIONS OF REINFORCED DUCT LINE ENCASEMENT. USE UNDER ALL RAILROAD TRACKS AND IN OTHER AREAS AS NEEDED. REINFORCED EN-CASEMENT TO EXTEND 15' MIN. ON EACH SIDE OF CENTERLINE OF RR TRACKS.

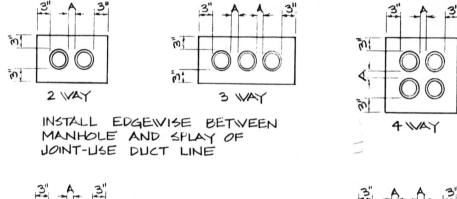

INSTALL EDGEWISE BETWEEN MANHOLE AND SPLAY OF JOINT-USE DUCT LINE

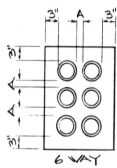

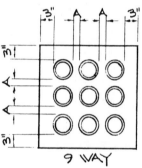

NOTES:
1. DIMENSION "A" IS 1½" FOR CLAY OR SOAPSTONE DUCT AND 2" FOR FIBER OR ASBESTOS CEMENT DUCT.
2. ALL DIMENSIONS ARE MINIMUM FROM OUTSIDE SURFACE OF DUCT.
3. CONCRETE NOT REQUIRED BETWEEN HORIZONTAL FACES OF CLAY DUCT.

## FIGURE 3-27

## TYPICAL SECTIONS OF DUCT LINES
### NO SCALE

---

Figure 3-27 is a set of details showing standard duct banks and the basic arrangements and clearances between ducts required in the duct bank. The details are relatively self-explanatory.

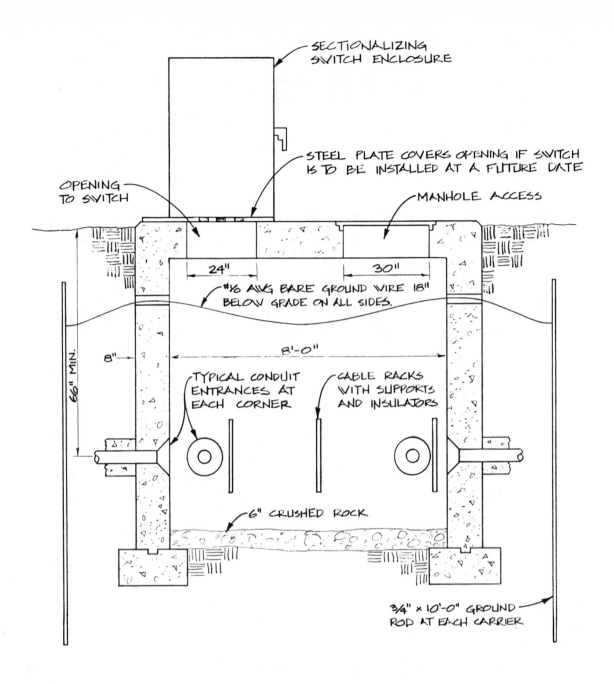

SECTIONALIZING SWITCH ENCLOSURE

STEEL PLATE COVERS OPENING IF SWITCH IS TO BE INSTALLED AT A FUTURE DATE

OPENING TO SWITCH

MANHOLE ACCESS

24"

30"

#1/0 AWG BARE GROUND WIRE 18" BELOW GRADE ON ALL SIDES.

66" MIN.

8"

8'-0"

TYPICAL CONDUIT ENTRANCES AT EACH CORNER

CABLE RACKS WITH SUPPORTS AND INSULATORS

6" CRUSHED ROCK

3/4" x 10'-0" GROUND ROD AT EACH CARRIER

**FIGURE 3-28**

SECTION THROUGH PADMOUNT SWITCH MANHOLE

NO SCALE

Figure 3-28 is a detail that is frequently used in high voltage work and shows a typical section through a pad mounted transformer switch manhole. The actual number of racks, the actual size of conduits, and other items would have to vary to suit the given installation.

274

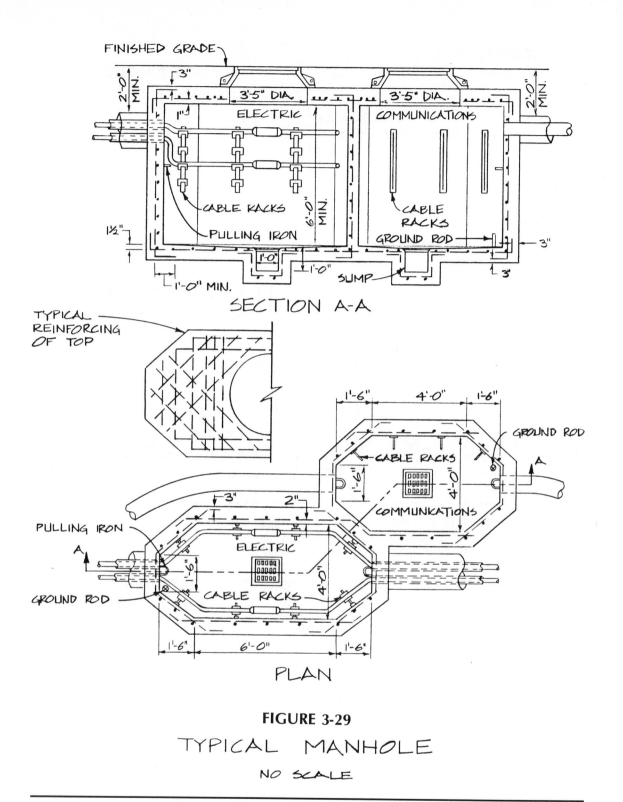

FINISHED GRADE

2'-0" MIN.

3"

3'-5" DIA.

3'-5" DIA.

2'-0" MIN.

1"

ELECTRIC

COMMUNICATIONS

6'-0" MIN.

CABLE RACKS

CABLE RACKS

PULLING IRON

GROUND ROD

3"

1½"

1'-0"

1'-0"

3'

1'-0" MIN.

SUMP

3"

SECTION A-A

TYPICAL
REINFORCING
OF TOP

1'-6"   4'-0"   1'-6"

GROUND ROD

CABLE RACKS

A

1'-6"

4'-0"

PULLING IRON

3"   2"

COMMUNICATIONS

A

ELECTRIC

1'-6"

4'-0"

GROUND ROD

CABLE RACKS

1'-6"   6'-0"   1'-6"

PLAN

**FIGURE 3-29**

TYPICAL MANHOLE

NO SCALE

Figure 3-29 illustrates a typical combination of electric and communication
manholes. There are many varieties of these used, both built-up and
prefabricated. When properly sized, these details would be typical for
most undergound distribution systems.

275

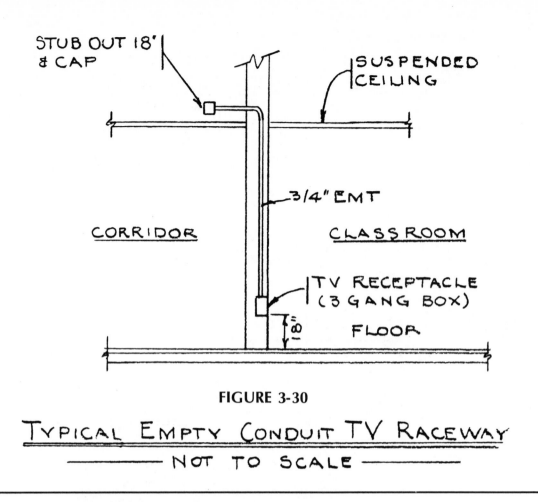

**FIGURE 3-30**

## TYPICAL EMPTY CONDUIT TV RACEWAY
### —— NOT TO SCALE ——

Figure 3-30 is a small detail of a commonly omitted or skipped item on a given set of construction drawings. It is the typical empty conduit TV detail, and the most important thing to note is that generally the TV receptacle is a three gang box.

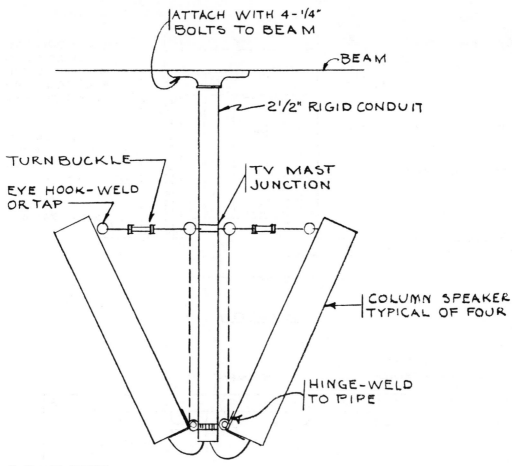

ATTACH WITH 4-1/4"
BOLTS TO BEAM

BEAM

2 1/2" RIGID CONDUIT

TURNBUCKLE

TV MAST
JUNCTION

EYE HOOK-WELD
OR TAP

COLUMN SPEAKER
TYPICAL OF FOUR

HINGE-WELD
TO PIPE

1/4" X 1" FLAT STEEL
DRILL & TAP IN SPEAKER

**FIGURE 3-31**

SOUND COLUMN SUPPORT DETAIL

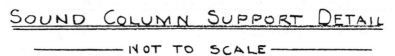

NOT TO SCALE

Figure 3-31 is one of many versions of a speaker column support, which is
frequently used in gymnasiums, halls, and similar areas. Speakers require
careful selection for each situation.

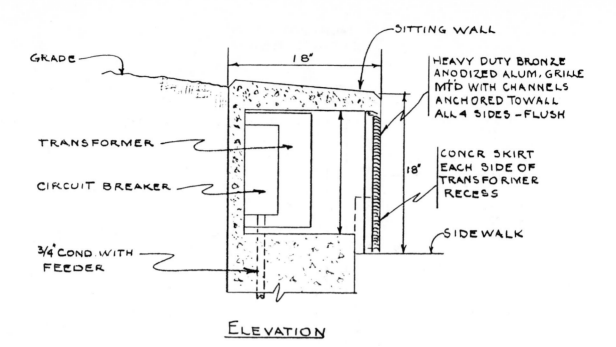

GRADE

TRANSFORMER

CIRCUIT BREAKER

3/4" COND. WITH FEEDER

—SITTING WALL

18"

HEAVY DUTY BRONZE ANODIZED ALUM. GRILLE MT'D WITH CHANNELS ANCHORED TO WALL ALL 4 SIDES—FLUSH

18"

CONCR SKIRT EACH SIDE OF TRANSFORMER RECESS

—SIDEWALK

ELEVATION

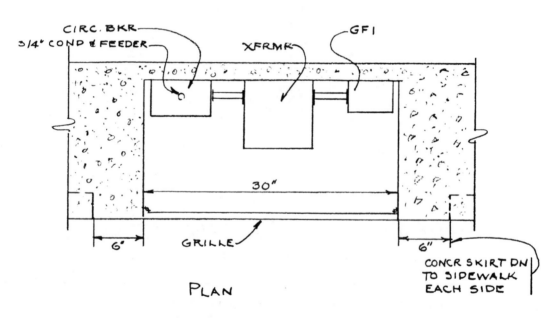

CIRC. BKR
3/4" COND & FEEDER

XFRMR

GFI

30"

6"

GRILLE

6"

CONCR SKIRT DN TO SIDEWALK EACH SIDE

PLAN

## FIGURE 3-32

## EXTERIOR TRANSFORMER WALL RECESSED

——————— NO SCALE ———————

Figure 3-32 is a typical detail of a small exterior transformer which might be recessed in a wall for an overall campus or other distribution requiring this type of installation. It is relatively standard and the only requirement would be to vary the dimensions to suit the given conditions.

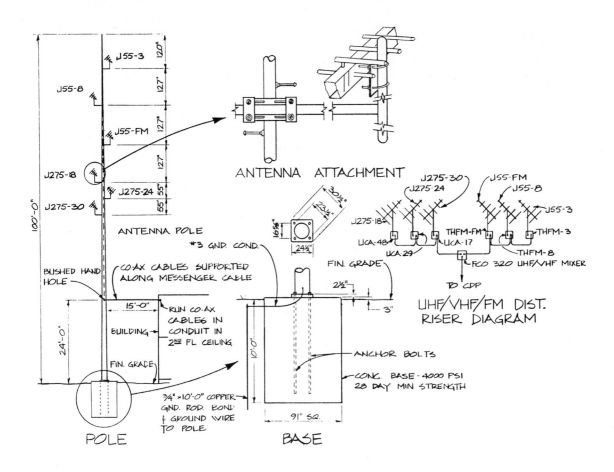

ANTENNA ATTACHMENT

UHF/VHF/FM DIST.
RISER DIAGRAM

POLE

BASE

**FIGURE 3-33**

MATV DETAILS
NO SCALE

Figure 3-33 is a master antenna TV pole detail, which is taken from an actual job. With modifications, it can be used on almost any job. The antenna noted is for a particular manufacturer. Changes in materials as selected by the engineer should be noted on the drawing.

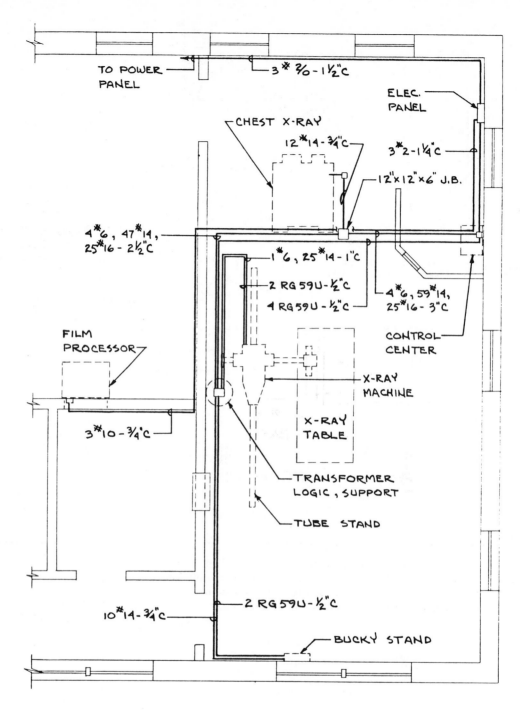

**FIGURE 3-34**

POWER WIRING OF X-RAY

Figure 3-34 is an illustration of a specialized retrofit changeover from one type of usage to another, where in an existing structure was switched to X-ray and the wiring was exposed. It would have to be modified to suit the equipment actually purchased for the job. The materials and sizes are not to be used as is.

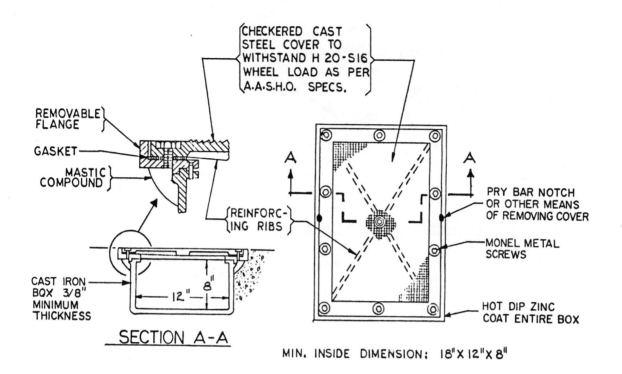

CHECKERED CAST STEEL COVER TO WITHSTAND H 20-S16 WHEEL LOAD AS PER A.A.S.H.O. SPECS.

REMOVABLE FLANGE

GASKET

MASTIC COMPOUND

REINFORCING RIBS

CAST IRON BOX 3/8" MINIMUM THICKNESS

12"

8"

SECTION A-A

A

A

PRY BAR NOTCH OR OTHER MEANS OF REMOVING COVER

MONEL METAL SCREWS

HOT DIP ZINC COAT ENTIRE BOX

MIN. INSIDE DIMENSION: 18" X 12" X 8"

FIGURE 3-35

WATERTIGHT SIGNAL
JUNCTION BOX
SCALE : NONE

Figure 3-35 is a standard, water type signal junction box.

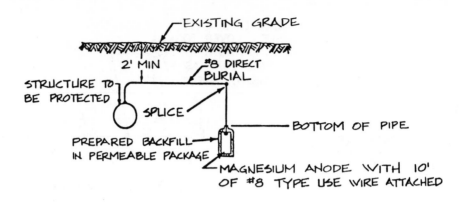

MAGNESIUM ANODE INSTALLATION

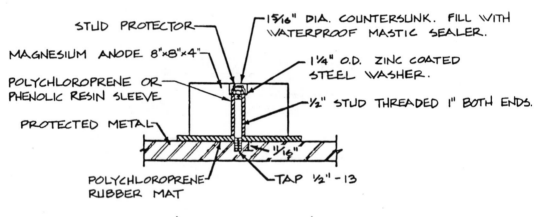

MAGNESIUM BLOCK ANODE
INSTALLATION FOR TANKS

FIGURE 3-36

CATHODIC PROTECTION

---

Figures 3-36 through 3-43 are typical cathodic protection details. Generally, these details are used in various combinations, depending on the type of anode and cathodic protection designed by the corrosion engineer. Frequently, this area of design is so specialized that the firm doing this type of work would be well advised to use a specialist in corrosion type work as a consultant. Without soil testing in advance of the work and total calculation of the amperage, voltage, ground resistance, etc., required, the details are of very little value. Their value becomes apparent in that after these are selected, one or more of these details may easily apply.

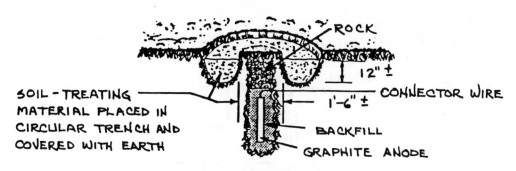

SOIL-TREATING
MATERIAL PLACED IN
CIRCULAR TRENCH AND
COVERED WITH EARTH

ROCK

12" ±

CONNECTOR WIRE

1'-6" ±

BACKFILL

GRAPHITE ANODE

TRENCH METHOD

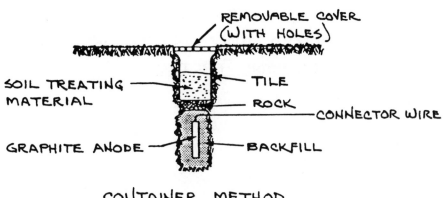

REMOVABLE COVER
(WITH HOLES)

SOIL TREATING
MATERIAL

TILE

ROCK

CONNECTOR WIRE

GRAPHITE ANODE

BACKFILL

CONTAINER METHOD

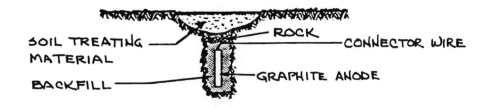

SOIL TREATING
MATERIAL

ROCK

CONNECTOR WIRE

BACKFILL

GRAPHITE ANODE

BASIN METHOD

FIGURE 3-37

VARIOUS SOIL TREATMENTS FOR
GRAPHITE ANODE GROUND ROD
CATHODIC PROTECTION

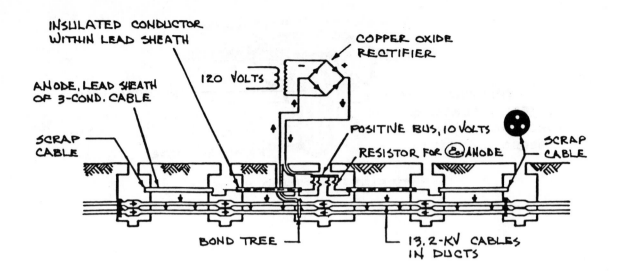

INSULATED CONDUCTOR
WITHIN LEAD SHEATH

COPPER OXIDE
RECTIFIER

120 VOLTS

ANODE, LEAD SHEATH
OF 3-COND. CABLE

POSITIVE BUS, 10 VOLTS

SCRAP
CABLE

RESISTOR FOR (Co) ANODE

SCRAP
CABLE

BOND TREE

13.2-KV CABLES
IN DUCTS

FIGURE 3-38

SECTIONALIZED ANODES IN CABLE INSTALLATION

CATHODIC PROTECTION

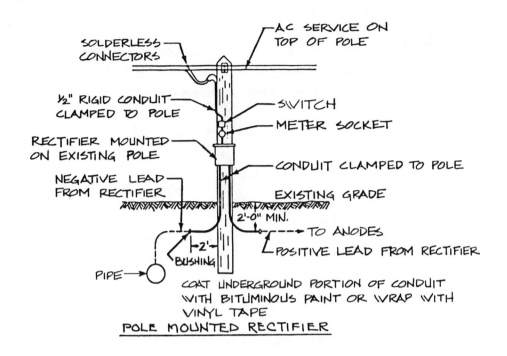

SOLDERLESS CONNECTORS

AC SERVICE ON TOP OF POLE

½" RIGID CONDUIT CLAMPED TO POLE

SWITCH

METER SOCKET

RECTIFIER MOUNTED ON EXISTING POLE

CONDUIT CLAMPED TO POLE

NEGATIVE LEAD FROM RECTIFIER

EXISTING GRADE

2'-0" MIN.

TO ANODES

POSITIVE LEAD FROM RECTIFIER

2'

BUSHING

PIPE

COAT UNDERGROUND PORTION OF CONDUIT WITH BITUMINOUS PAINT OR WRAP WITH VINYL TAPE

POLE MOUNTED RECTIFIER

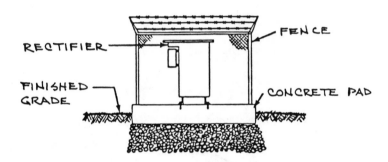

RECTIFIER

FENCE

FINISHED GRADE

CONCRETE PAD

ENCLOSED RECTIFIER PAD INSTALLATION

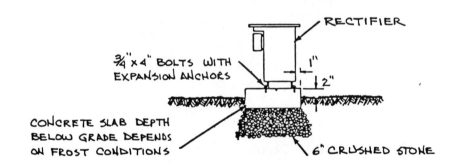

RECTIFIER

¾" x 4" BOLTS WITH EXPANSION ANCHORS

1"

2"

CONCRETE SLAB DEPTH BELOW GRADE DEPENDS ON FROST CONDITIONS

6" CRUSHED STONE

**FIGURE 3-39**

RECTIFIER PAD INSTALLATION

TYPICAL RECTIFIER INSTALLATIONS

285

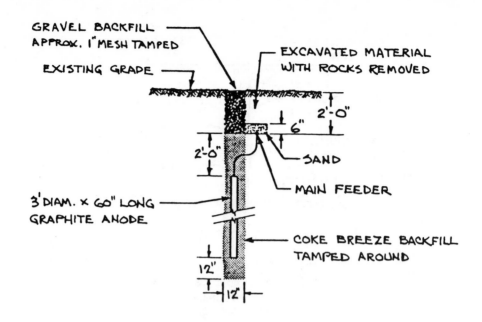

GRAVEL BACKFILL
APPROX. 1" MESH TAMPED

EXISTING GRADE

EXCAVATED MATERIAL
WITH ROCKS REMOVED

2'-0"

6"

SAND

2'-0"

MAIN FEEDER

3' DIAM. x 60" LONG
GRAPHITE ANODE

COKE BREEZE BACKFILL
TAMPED AROUND

12"

12"

VERTICAL GRAPHITE ANODE

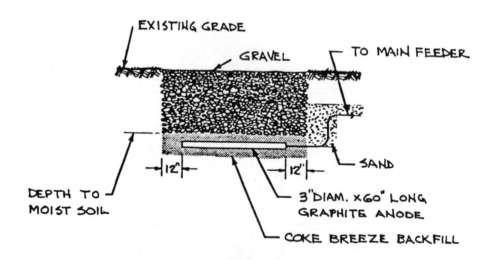

EXISTING GRADE

GRAVEL

TO MAIN FEEDER

DEPTH TO
MOIST SOIL

12"

12"

SAND

3" DIAM. x 60" LONG
GRAPHITE ANODE

COKE BREEZE BACKFILL

HORIZONTAL GRAPITE ANODE

FIGURE 3-40

TYPICAL GRAPHITE ANODE INSTALLATIONS
FOR CATHODIC PROTECTION

286

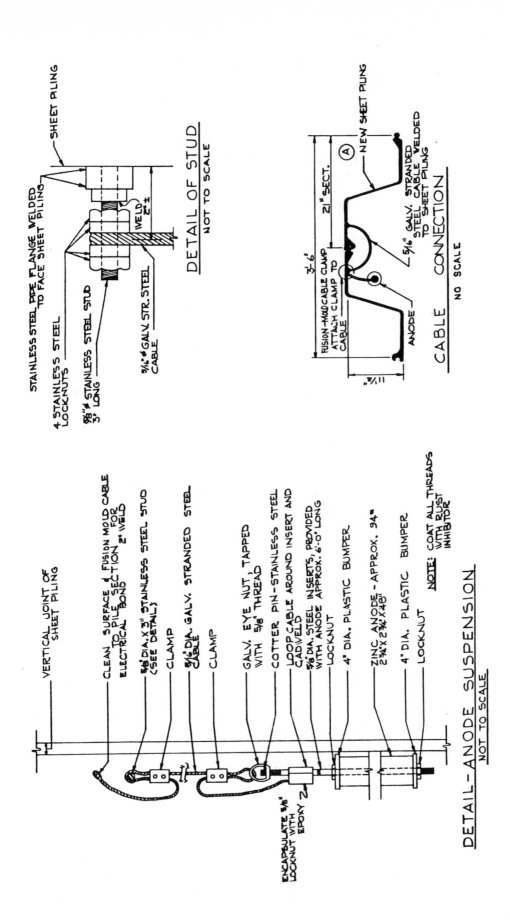

DETAIL OF STUD
NOT TO SCALE

SHEET PILING

STAINLESS STEEL PIPE FLANGE WELDED TO FACE SHEET PILING

4 STAINLESS STEEL LOCKNUTS

WELD 2"±

⅝"ø STAINLESS STEEL STUD 3" LONG

5/16"ø GALV. STR. STEEL CABLE

CABLE CONNECTION
NO SCALE

NEW SHEET PILING

5/16"ø GALV. STRANDED STEEL CABLE WELDED TO SHEET PILING

FUSION-MOLD CABLE CLAMP ATTACH CLAMP TO CABLE

ANODE

21" SECT.

3'-6"

11 ½"

Ⓐ

DETAIL-ANODE SUSPENSION
NOT TO SCALE

VERTICAL JOINT OF SHEET PILING

CLEAN SURFACE & FUSION MOLD CABLE FOR ELECTRICAL BOND TO PILE SECTION 2" WELD

⅝"DIA.X3" STAINLESS STEEL STUD (SEE DETAIL)

CLAMP

5/16"DIA. GALV. STRANDED STEEL CABLE

CLAMP

GALV. EYE NUT, TAPPED WITH ⅝" THREAD

COTTER PIN-STAINLESS STEEL

LOOP CABLE AROUND INSERT AND CADWELD

⅝"DIA. STEEL INSERTS, PROVIDED WITH ANODE APPROX. 6'-0" LONG

LOCKNUT

4" DIA. PLASTIC BUMPER

ZINC ANODE - APPROX. 94# 2¾"x2¾"x48"

4" DIA. PLASTIC BUMPER

LOCKNUT

ENCAPSULATE ⅝" LOCKNUT WITH EPOXY 2"

NOTE: COAT ALL THREADS WITH RUST INHIBITOR

FIGURE 3-41

QUAY WALL SHEET PILING CATHODIC PROTECTION

287

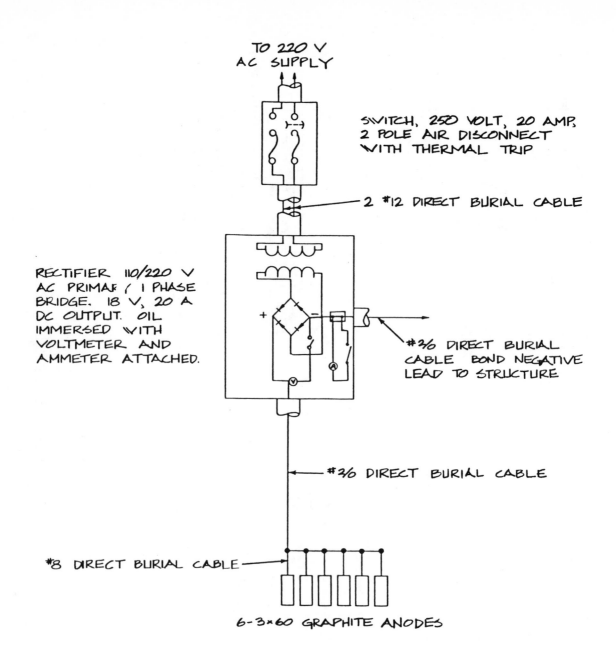

TO 220 V
AC SUPPLY

SWITCH, 250 VOLT, 20 AMP,
2 POLE AIR DISCONNECT
WITH THERMAL TRIP

2 #12 DIRECT BURIAL CABLE

RECTIFIER 110/220 V
AC PRIMARY 1 PHASE
BRIDGE. 18 V, 20 A
DC OUTPUT. OIL
IMMERSED WITH
VOLTMETER AND
AMMETER ATTACHED.

#3/0 DIRECT BURIAL
CABLE BOND NEGATIVE
LEAD TO STRUCTURE

#3/0 DIRECT BURIAL CABLE

#8 DIRECT BURIAL CABLE

6-3×60 GRAPHITE ANODES

**FIGURE 3-42**

ELECTRICAL WIRING DIAGRAM
FOR THE IMPRESSED CURRENT SYSTEM

NO SCALE

288

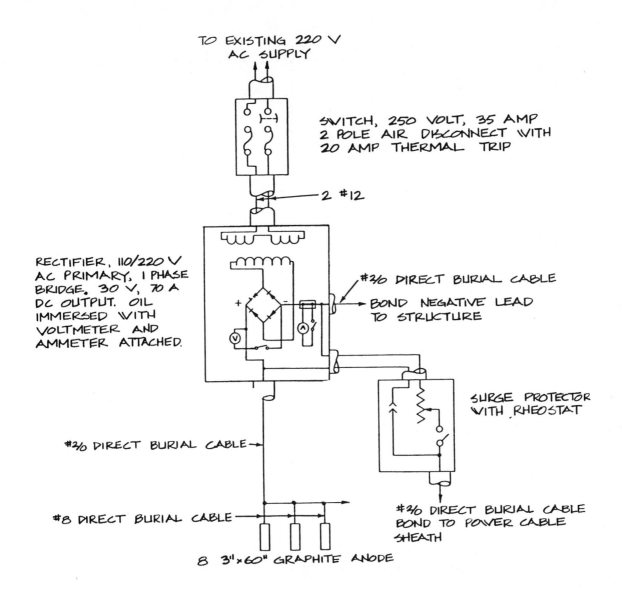

TO EXISTING 220 V
AC SUPPLY

SWITCH, 250 VOLT, 35 AMP
2 POLE AIR DISCONNECT WITH
20 AMP THERMAL TRIP

2 #12

RECTIFIER, 110/220 V
AC PRIMARY, 1 PHASE
BRIDGE. 30 V, 70 A
DC OUTPUT. OIL
IMMERSED WITH
VOLTMETER AND
AMMETER ATTACHED.

#3/0 DIRECT BURIAL CABLE
BOND NEGATIVE LEAD
TO STRUCTURE

SURGE PROTECTOR
WITH RHEOSTAT

#3/0 DIRECT BURIAL CABLE →

#8 DIRECT BURIAL CABLE →

#3/0 DIRECT BURIAL CABLE
BOND TO POWER CABLE
SHEATH

8 3"x60" GRAPHITE ANODE

**FIGURE 3-43**

ELECTRICAL WIRING DIAGRAM FOR
IMPRESSED CURRENT SYSTEM, WITH SURGE PROTECTOR
NO SCALE

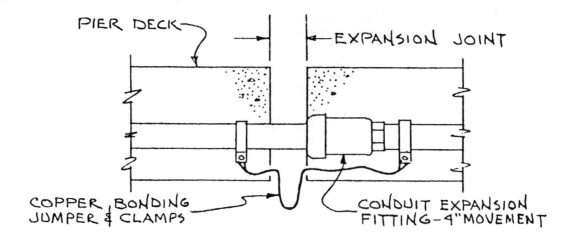

ELEVATION

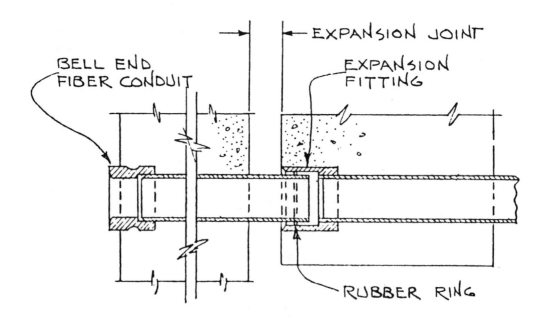

SECTION

**FIGURE 3-44**

PIER DECK CONDUIT EXPANSION CONNECTION

Figure 3-44 is a typical expansion joint in a pier or dock. It is important to note that a bonding jumper is required to maintain integrity of the conduit as a ground.

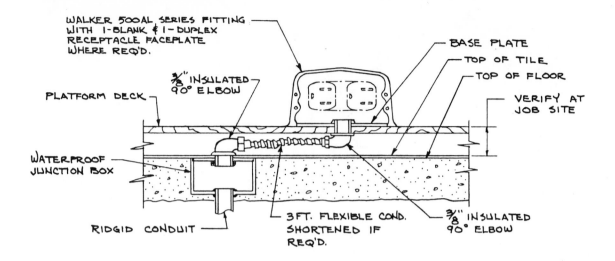

**FIGURE 3-45**

RAISED PLATFORM CONVENIENCE OUTLET

Figure 3-45 is relatively self-explanatory and is based on using a specific manufacturer. It indicates the use of a convenience outlet on a raised platform, which is a common retrofit requirement.

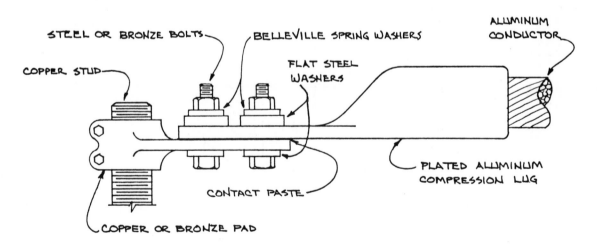

**FIGURE 3-46**

CONNECTION OF LARGE ALUMINUM CONDUCTORS
TO COPPER STUDS OR COPPER TERMINAL PADS

Figure 3-46 is an enlargement of a detail that may be used when aluminum conductors are connected to copper conductors.

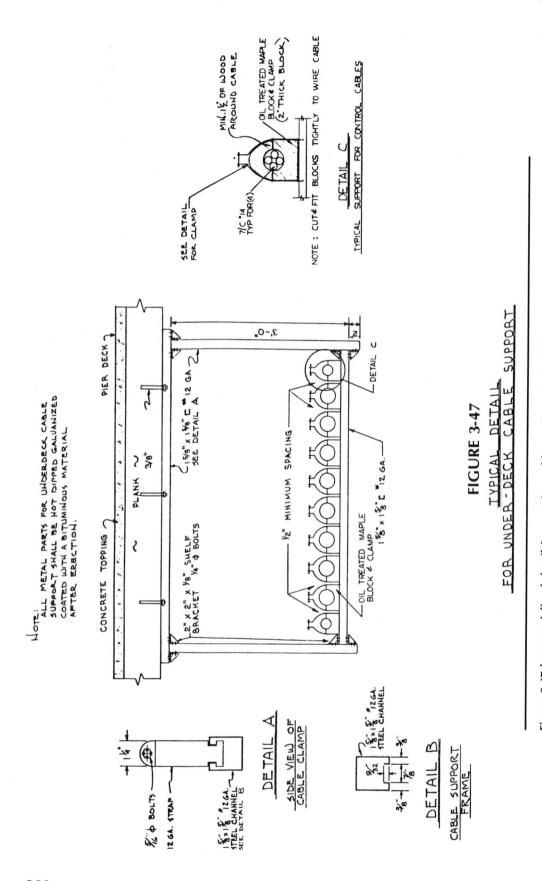

**FIGURE 3-47**

TYPICAL DETAIL
FOR UNDER-DECK CABLE SUPPORT

Figure 3-47 is a specialized detail for carrying cables under a dock. This detail was taken from a specific project. All items would require careful analysis and sizing by the engineer.

292

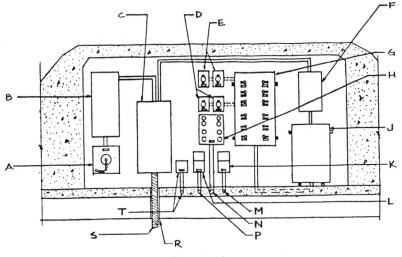

ELEVATION

A  WELDING OUTLET 200 A, 3 W, 4 POLE, 480 V BRASS WEATHER RESISTANT
     RAINTIGHT RECEPTACLE WITH GASKETED SCREW COVER.
B  INTEGRALLY FUSED CIRCUIT BREAKER 200 A, 3 ∅, 480 V, 100,000 A.I.C.
     CLASS 2C (FED. SPEC. WC 375a) IN WATERTIGHT ENCLOSURE NEMA TYPE 4
     FOR WELDING OUTLET.
C  CAST METAL GASKETED WATERTIGHT ENCLOSURE (JB) 12" W × 24" H × 15" D.
D  UTILITY OUTLETS: 2-20 A, 4 POLE, 3 ∅, 208 V GND TYPE WITH CAST SPRING
     LOADED COVER PLATE GASKETED AND MOUNTED ON CAST METAL
     WATERTIGHT BOX.
E  UTILITY OUTLETS: 2 DUPLEX RECEPTACLE OUTLETS 20 A, 125 V GND TYPE
     EACH WITH BOX AND COVER PLATE AS NOTED IN "D" ABOVE.
F  TRANSFORMER PRIMARY CIRCUIT BREAKER 30 A, 3 P, 480 V, 100,000 A.I.C.,
     CLASS 2f (FED SPEC WC 375a) IN WATERTIGHT ENCLOSURE NEMA TYPE 4.
G  PANEL.
H  TELEPHONE OUTLET, 12" × 8" × 4" CAST METAL GASKETED WATERTIGHT BOX.
     GOVT SHALL FURNISH TELEPHONE TYPE JACK TO CONTR. FOR INSTALLATION.
J  OUTDOOR DRY TYPE TRANSFORMER 15 KVA, 480 V/208/120 V, 3 ∅, 4 W, 60 HZ.
     ENCAPSULATED WINDING.
K  TELEVISION ANTENNA SYSTEM OUTLET 2-4"×4"×4" CAST METAL GASKETED
     WATERTIGHT BOXES.
L  FIRE ALARM AUXILIARY JACK OUTLET 2-4"×4"×4" CAST METAL GASKETED
     WATERTIGHT BOXES WITH SPRING LOADED COVER – PROVIDE 2-10 A,
     2 P, 125 V RECEPTACLES AND MATCHING PLUGS (THREADED TYPE).
M  1" TELEVISION CONDUIT.
N  2" TELEPHONE CONDUIT.
P  2 #12 IN 1" CONDUIT FOR FIRE ALARM.
R  WEATHERPROOF SLEEVE.
S  3/C #500 MCM.
T  SPARE CONDUIT AND OUTLET BOX (SAME AS TV).


# FIGURE 3-48

TYPICAL PIER POWER CENTER
IN CONCRETE HOOD ENCLOSURE

NOT TO SCALE

Figure 3-48 shows the typical arrangement of a dock power center
enclosed in a concrete hood. This particular power center was taken from
a particular job, and the sizes shown are pertinent to that job. Obviously,
the engineer involved would have to design more of the system to suit a
given installation. Finally, in a pier or dock, the sleeve and cable termination installation frequently must be detailed.

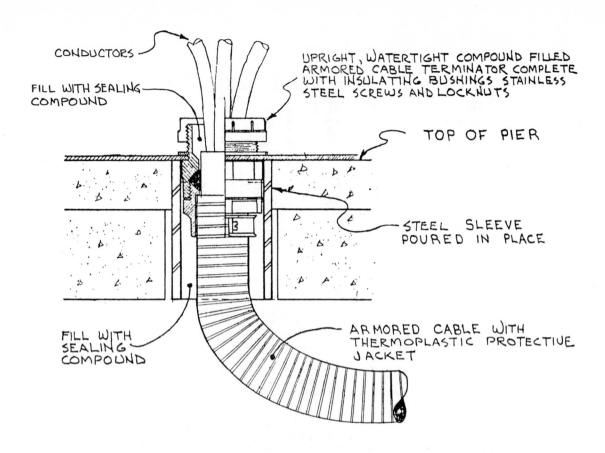

CONDUCTORS

FILL WITH SEALING COMPOUND

UPRIGHT, WATERTIGHT COMPOUND FILLED ARMORED CABLE TERMINATOR COMPLETE WITH INSULATING BUSHINGS STAINLESS STEEL SCREWS AND LOCKNUTS

TOP OF PIER

STEEL SLEEVE POURED IN PLACE

FILL WITH SEALING COMPOUND

ARMORED CABLE WITH THERMOPLASTIC PROTECTIVE JACKET

**FIGURE 3-49**

TYPICAL PIER SLEEVE AND CABLE TERMINATION

Figure 3-49 is a standard detail of a pier sleeve and cable termination. Where applicable, this detail can be traced as is. The steel sleeve should preferably be a fairly heavy gauge.

294

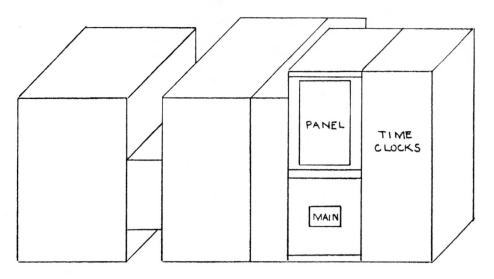

SITE LIGHTING SUB STATION — EXTERIOR VIEW

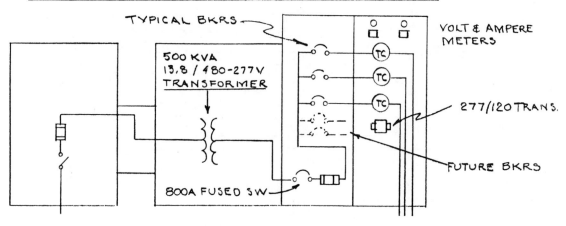

**FIGURE 3-50**

SITE LIGHTING SUB-STATION — ONE-LINE DIAGRAM

Figure 3-50 is a typical site lighting substation, or a substation for anything, and is merely an indication of how a substation might appear. Panels, transformers, and sizes of all equipment would require adjustment to suit the particular project.

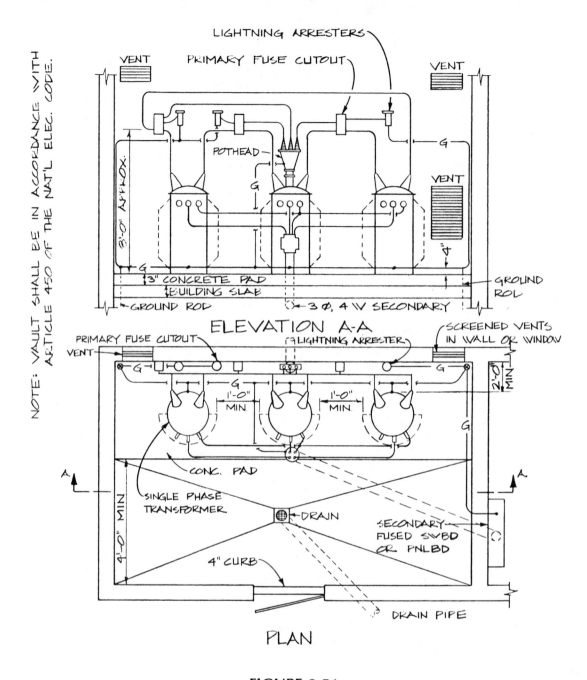

**FIGURE 3-51**

TYPICAL INDOOR CONVENTIONAL
TRANSFORMER INSTALLATION

NO SCALE

Figure 3-51 is an indoor version of the outdoor transformer stations that
have been shown previously. The vent should be carefully sized according to local and national codes, and the curb at the door is normally a
requirement.

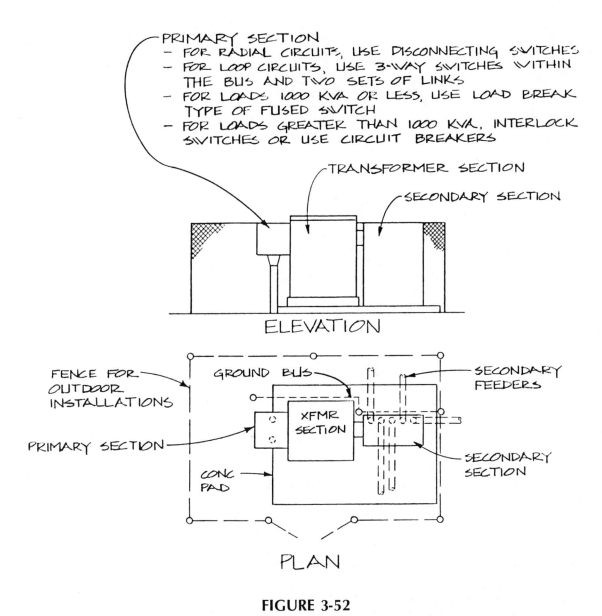

PRIMARY SECTION
- FOR RADIAL CIRCUITS, USE DISCONNECTING SWITCHES
- FOR LOOP CIRCUITS, USE 3-WAY SWITCHES WITHIN THE BUS AND TWO SETS OF LINKS
- FOR LOADS 1000 KVA OR LESS, USE LOAD BREAK TYPE OF FUSED SWITCH
- FOR LOADS GREATER THAN 1000 KVA, INTERLOCK SWITCHES OR USE CIRCUIT BREAKERS

TRANSFORMER SECTION

SECONDARY SECTION

ELEVATION

FENCE FOR OUTDOOR INSTALLATIONS

GROUND BUS

SECONDARY FEEDERS

XFMR SECTION

PRIMARY SECTION

CONC PAD

SECONDARY SECTION

PLAN

**FIGURE 3-52**

TYPICAL LIQUID-COOLED
TRANSFORMER LOAD CENTER

NO SCALE

Figure 3-52 indicates a typical liquid cooled transformer load center. This drawing is relatively simple. The fence and its clearances should be checked against local codes, as well as the National Electric Code.

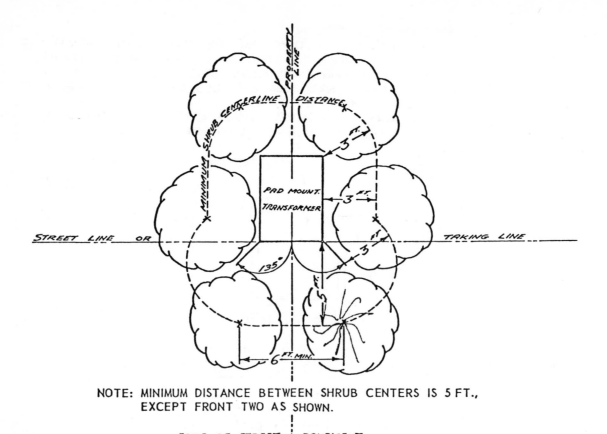

NOTE: MINIMUM DISTANCE BETWEEN SHRUB CENTERS IS 5 FT., EXCEPT FRONT TWO AS SHOWN.

EDGE OF STREET PAVEMENT

## FIGURE 3-53

## SHRUBBERY SCREEN
## FOR PAD MOUNTED TRANSFORMER

Figures 3-53 through 3-57 are the details that customarily go with an exterior pad mounted transformer installation. This series of details covers shrubbery around the installation, the pad for the installation, the various and sundry drains, the grounding connections, the openings in the mat, the thickness of the mat, and the typical installation. These dimensions would have to be varied for the given installation, but this series of details is an excellent standard for any situation that the engineer might encounter.

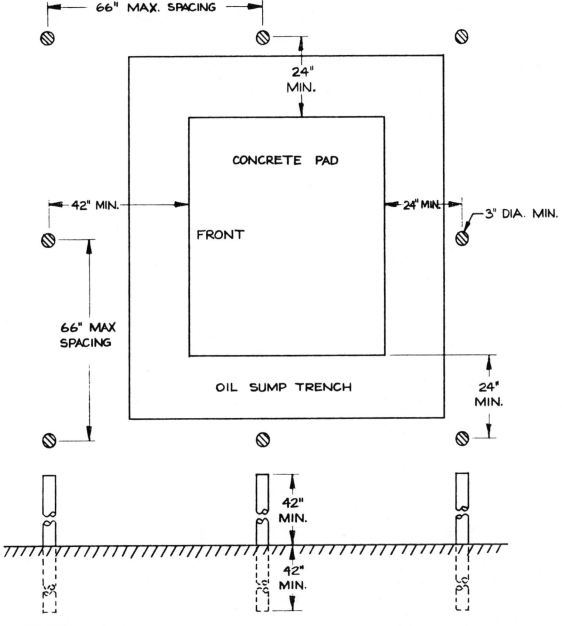

NOTES:
1. PROTECTIVE BUMPERS ARE TO BE INSTALLED ON SIDES EXPOSED TO VEHICULAR TRAFFIC, ONLY.
2. STEEL PIPES 3" MINIMUM FILLED WITH CONCRETE, I-BEAMS 5" MINIMUM OR OTHER SUITABLE MEANS OF PROTECTION MAY BE USED. HEAVIER BUMPERS SET DEEPER SHOULD BE CONSIDERED WHERE EXPOSED TO HEAVY TRUCKS.

**FIGURE 3-54**

PAD MOUNTED TRANSFORMER OR TRANSCLOSURE
PROTECTIVE BUMPER INSTALLATION
NO SCALE

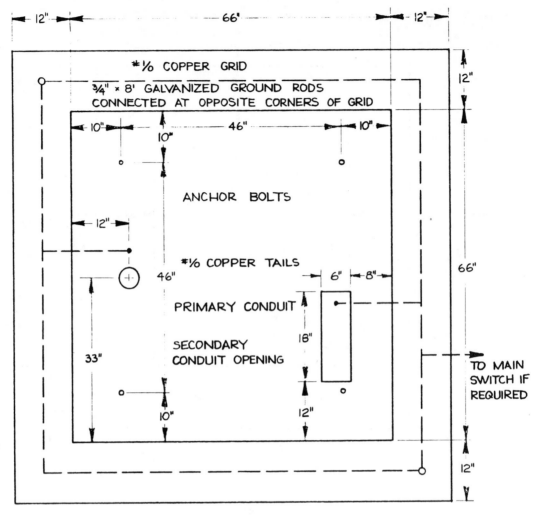

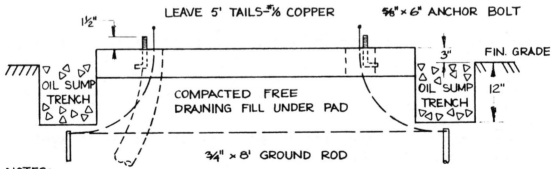

NOTES:
1. CONCRETE PAD TO BE 36" MINIMUM FROM WALL, FENCE OR OTHER OBSTRUCTION.
2. EXACT LOCATION OF ANCHOR BOLTS AND PRIMARY & SECONDARY CONDUIT OPENINGS TO BE DETERMINED FROM FINAL APPROVED SHOP DRAWINGS

**FIGURE 3-55**

TRANSCLOSURE PAD INSTALLATION
4.16 KV & 4.8 KV 3 Ø 150 KVA MAX.
NO SCALE

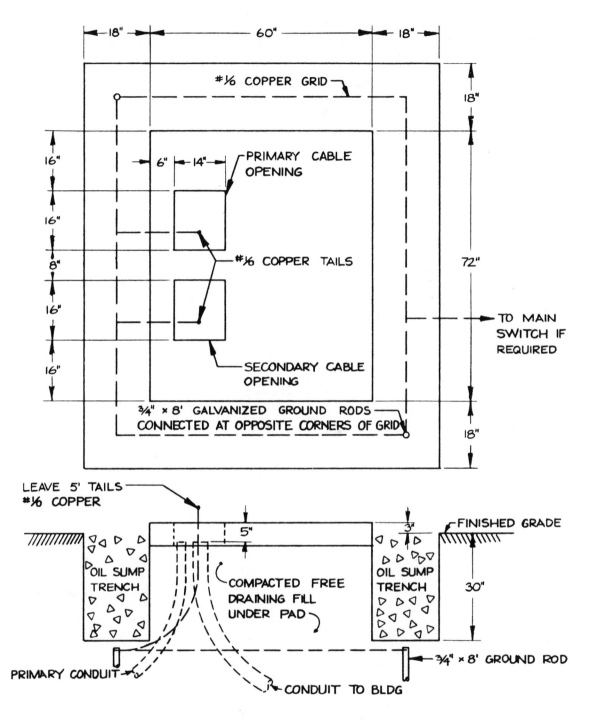

NOTE: CONCRETE PAD TO BE 36" MINIMUM FROM WALL, FENCE OR OTHER OBSTRUCTION

FIGURE 3-56

PAD MOUNT TRANSFORMER INSTALLATION
2.4 - 23 KV SYSTEMS 3 ∅ 500 KVA MAX
NO SCALE

301

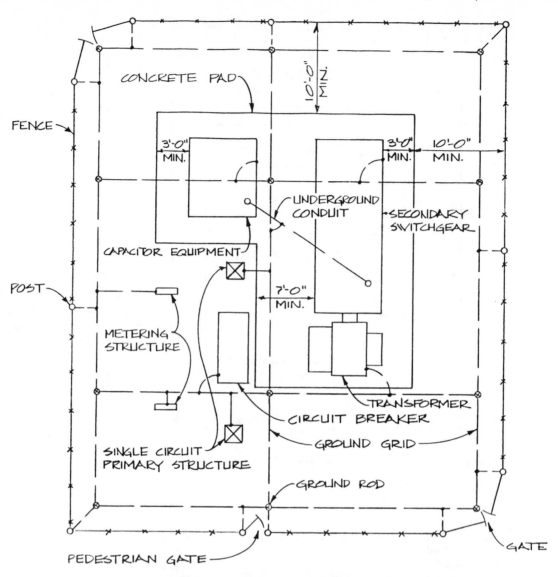

NOTES: GROUND EACH COLUMN OF METERING & PRIMARY STRUCTURES,
GROUND ALL METALLIC CONDUITS, ELEC. EQPT & GROUND BUS
GROUND FENCE POSTS AT MIDPOINTS AND CORNERS, AND ALL
GATE POSTS.

**FIGURE 3-57**

TYPICAL GROUNDING PLAN
PRIMARY UNIT SUBSTATION
NO SCALE

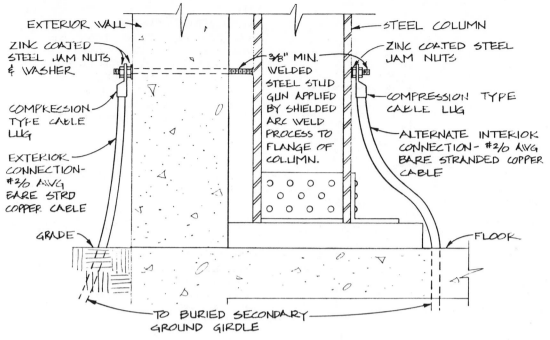

METHOD FOR GROUNDING STEEL COLUMNS

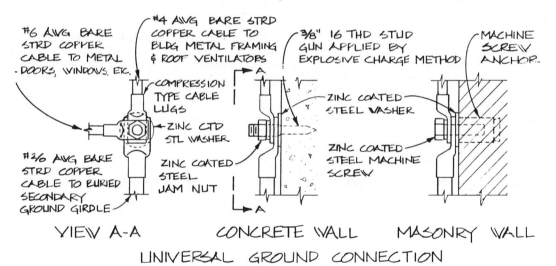

VIEW A-A     CONCRETE WALL     MASONRY WALL

UNIVERSAL GROUND CONNECTION

FIGURE 3-58

GROUNDING DETAILS FOR SECONDARY SYSTEM
NO SCALE

Figures 3-58 through 3-61 are typical ground connection details. These details may vary slightly with the latest code. No representation is given that they are exactly correct, but it is believed that they are currently relatively correct. They indicate the various grounding of steel; the grounding, for example, of a railroad track; and the connections to grounding electrodes. One or more of these details would be applicable on almost any job.

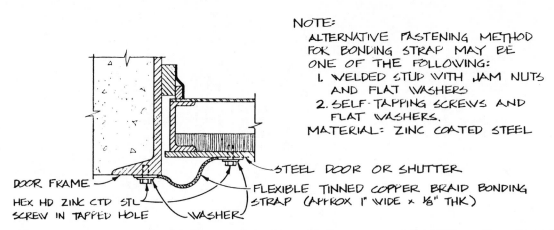

NOTE:
ALTERNATIVE FASTENING METHOD
FOR BONDING STRAP MAY BE
ONE OF THE FOLLOWING:
1. WELDED STUD WITH JAM NUTS
AND FLAT WASHERS
2. SELF-TAPPING SCREWS AND
FLAT WASHERS.
MATERIAL: ZINC COATED STEEL

STEEL DOOR OR SHUTTER

DOOR FRAME

FLEXIBLE TINNED COPPER BRAID BONDING
STRAP (APPROX 1" WIDE × ⅛" THK)

HEX HD ZINC CTD STL
SCREW IN TAPPED HOLE          WASHER

## METHOD FOR BONDING STEEL DOOR OR SHUTTER TO FRAME

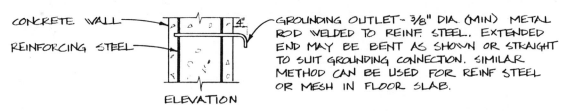

CONCRETE WALL

REINFORCING STEEL

GROUNDING OUTLET- ⅜" DIA. (MIN) METAL
ROD WELDED TO REINF. STEEL. EXTENDED
END MAY BE BENT AS SHOWN OR STRAIGHT
TO SUIT GROUNDING CONNECTION. SIMILAR
METHOD CAN BE USED FOR REINF STEEL
OR MESH IN FLOOR SLAB.

ELEVATION

## METHOD FOR GROUNDING OF REINFORCING STEEL

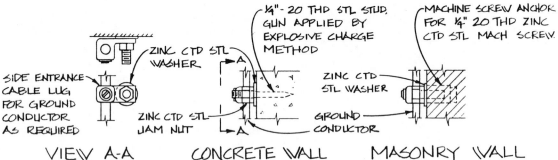

ZINC CTD STL
WASHER

SIDE ENTRANCE
CABLE LUG
FOR GROUND
CONDUCTOR
AS REQUIRED

ZINC CTD STL
JAM NUT

¼"- 20 THD STL STUD,
GUN APPLIED BY
EXPLOSIVE CHARGE
METHOD

MACHINE SCREW ANCHOR
FOR ¼" 20 THD ZINC
CTD STL MACH SCREW.

ZINC CTD
STL WASHER

GROUND
CONDUCTOR

VIEW A-A          CONCRETE WALL          MASONRY WALL

## METHOD FOR SUPPORTING GROUND CONDUCTORS ON WALLS

**FIGURE 3-59**

## GROUNDING DETAILS FOR SECONDARY SYSTEM

NO SCALE

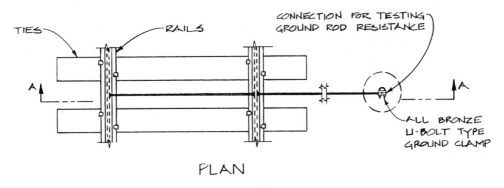

TIES

RAILS

CONNECTION FOR TESTING
GROUND ROD RESISTANCE

A

A

ALL BRONZE
U-BOLT TYPE
GROUND CLAMP

PLAN

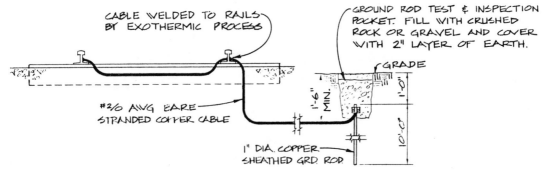

CABLE WELDED TO RAILS
BY EXOTHERMIC PROCESS

GROUND ROD TEST & INSPECTION
POCKET. FILL WITH CRUSHED
ROCK OR GRAVEL AND COVER
WITH 2" LAYER OF EARTH.

GRADE

1'-6" MIN.

1'-0"

10'-0"

#2/0 AWG BARE
STRANDED COPPER CABLE

1" DIA. COPPER
SHEATHED GRD. ROD.

SECTION A-A

**FIGURE 3-60**

BONDING & GROUNDING OF RAILROAD TRACK

NO SCALE

305

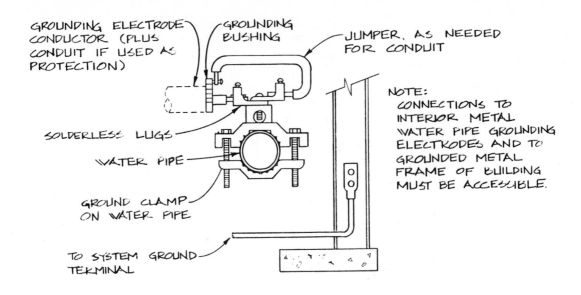

GROUNDING ELECTRODE CONDUCTOR (PLUS CONDUIT IF USED AS PROTECTION)

GROUNDING BUSHING

JUMPER, AS NEEDED FOR CONDUIT

SOLDERLESS LUGS

WATER PIPE

GROUND CLAMP ON WATER PIPE

TO SYSTEM GROUND TERMINAL

NOTE:
CONNECTIONS TO INTERIOR METAL WATER PIPE GROUNDING ELECTRODES AND TO GROUNDED METAL FRAME OF BUILDING MUST BE ACCESSIBLE.

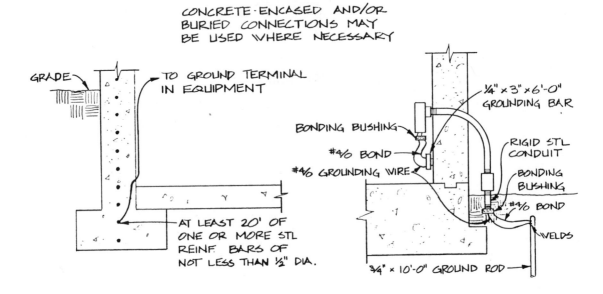

CONCRETE-ENCASED AND/OR BURIED CONNECTIONS MAY BE USED WHERE NECESSARY

GRADE

TO GROUND TERMINAL IN EQUIPMENT

AT LEAST 20' OF ONE OR MORE STL REINF. BARS OF NOT LESS THAN ½" DIA.

¼" × 3" × 6'-0" GROUNDING BAR

BONDING BUSHING

#4/0 BOND

#4/0 GROUNDING WIRE

RIGID STL CONDUIT

BONDING BUSHING

#4/0 BOND

WELDS

¾" × 10'-0" GROUND ROD

**FIGURE 3-61**

CONNECTIONS TO GROUNDING ELECTRODES

NO SCALE

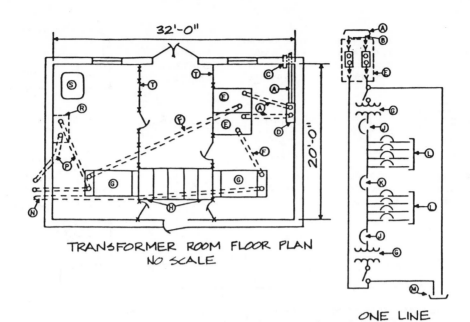

TRANSFORMER ROOM FLOOR PLAN
NO SCALE

ONE LINE
WIRING DIAGRAM
NO SCALE

A   INCOMING SERVICE - TWO FEEDERS PREFERRED
B   TIE BUS - INSTALL WHEN ONLY ONE FEEDER
      IS AVAILABLE
C   SPARE DUCTS
D   PULL BOX
E   MAIN CIRCUIT BREAKERS
F   PRIMARY CABLE IN DUCT
G   THREE PHASE TRANSFORMER
H   SECONDARY SWITCHGEAR SECTION
J   SECONDARY BREAKERS WITH AUXILIARY ALARM CONTACTS
K   TIE BREAKER INTERLOCKED WITH MAIN BREAKERS
L   SECONDARY FEEDERS
M   CABLES TO OTHER LOAD CENTERS, IF REQUIRED
N   DUCTS FOR CABLES TO OTHER LOAD CENTERS, IF REQUIRED
P   ALTERNATE ROUTE OF ONE PRIMARY FEEDER FOR TAPS TO OTHER
      BUILDINGS, IF REQUIRED
R   JUNCTION BOX AND OIL FUSE CUTOUTS OR LOAD-BREAK FUSED
      DISCONNECTS, IF REQUIRED FOR TAP
S   STREET LIGHTING TRANSFORMER
T   REMOVABLE WIRE PARTITION

## FIGURE 3-62

## TYPICAL DOUBLE-ENDED TRANSFORMER LOAD CENTER–PRIMARY CIRCUIT BREAKERS

### NO SCALE

Figure 3-62 is a typical small, double ended substation. Each transformer and its associated equipment should be capable of carrying the essential loads of both sections. The arrangement shown does not permit the addition of unexpected future breakers. If this is required, alternate arrangements may be used. Both require more floor area and higher costs initially.

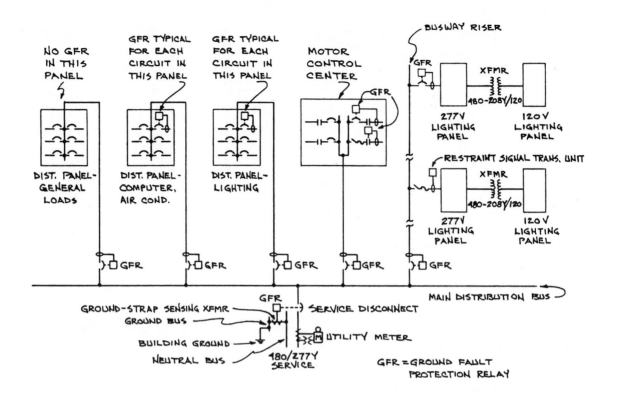

**FIGURE 3-63**

<u>SAMPLE APPLICATION OF</u>
<u>GROUND FAULT PROTECTION TO BUILDING ELECTRICAL SYSTEM</u>

Figure 3-63 is a drawing that is used as a guide for the application of ground fault protection. The ground fault protection, where required, should be incorporated in the schematic in a manner similar to that shown.

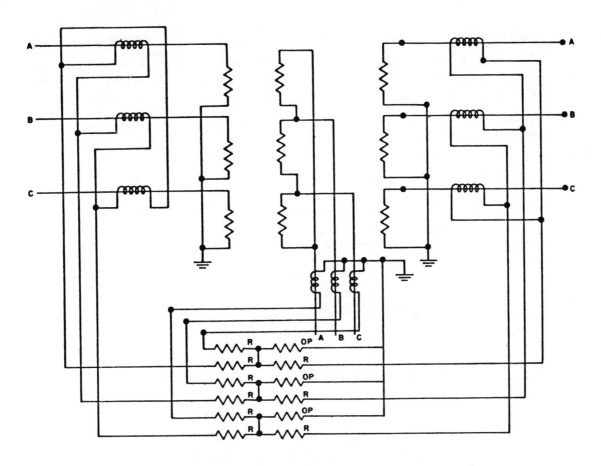

**FIGURE 3-64**

<u>CONNECTIONS FOR A</u>

<u>PERCENTAGE DIFFERENTIAL PROTECTIVE SYSTEM</u>

<u>FOR A THREE-WINDING TRANSFORMER</u>

Figures 3-64 and 3-65 are schematic drawings not to scale, which merely show the arrangement of a differential protective system for a three-winding and a two-winding transformer. The system, of course, would have to be properly sized and all of the components properly noted. The drawings are only to provide a guide for the designer.

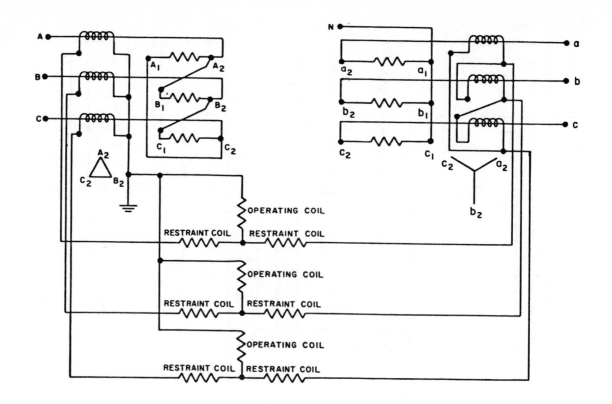

FIGURE 3-65

CONNECTIONS FOR A
PERCENTAGE DIFFERENTIAL PROTECTIVE SYSTEM
FOR A TWO-WINDING DELTA-STAR TRANSFORMER

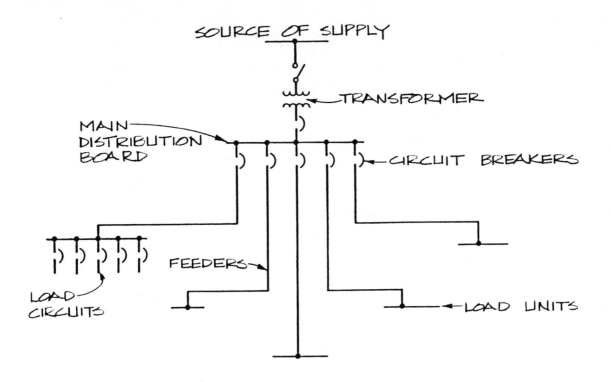

FIGURE 3-66

CONVENTIONAL SIMPLE-RADIAL
DISTRIBUTION SYSTEM

NO SCALE

Figures 3-66 through 3-72 are schematic drawings of the generally used distribution systems for 15,000-V and less systems. There are, of course, many systems other than the ones shown. The ones shown may not fit your installation exactly, but with some alteration they will fit. The systems shown are the ones generally used and include the simple radial system, the loop primary radial, the primary selective network, the primary selective spot network, the simple network, the primary selective radial, and the secondary selective radial. Regardless of the number of components, these details can be adjusted to the correct number of load centers that exist on your project, and with adjustment these schematic arrangements can be used in the schematic presentation of your system.

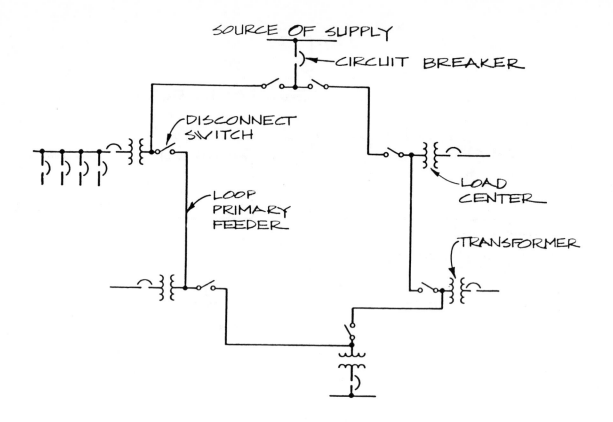

**FIGURE 3-67**

LOOP PRIMARY-RADIAL
DISTRIBUTION SYSTEM

NO SCALE

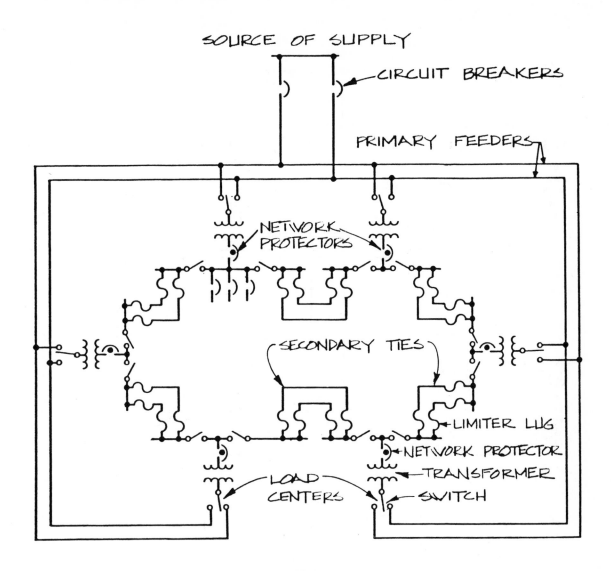

**FIGURE 3-68**

PRIMARY SELECTIVE NETWORK
DISTRIBUTION SYSTEM

NO SCALE

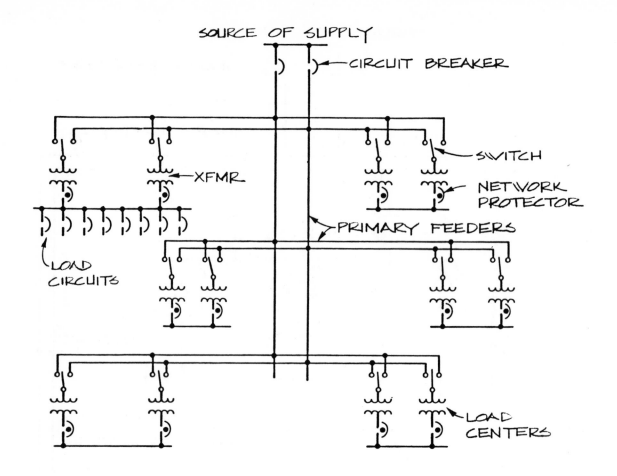

**FIGURE 3-69**

PRIMARY SELECTIVE SPOT-NETWORK
DISTRIBUTION SYSTEM

NO SCALE

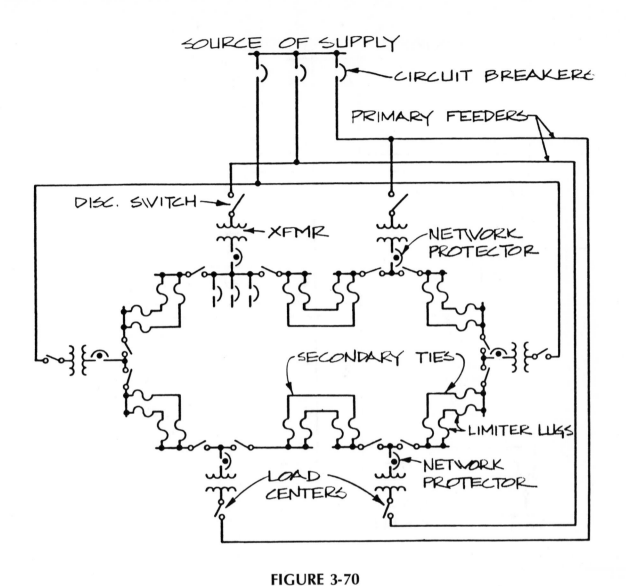

**FIGURE 3-70**

SIMPLE NETWORK
DISTRIBUTION SYSTEM

NO SCALE

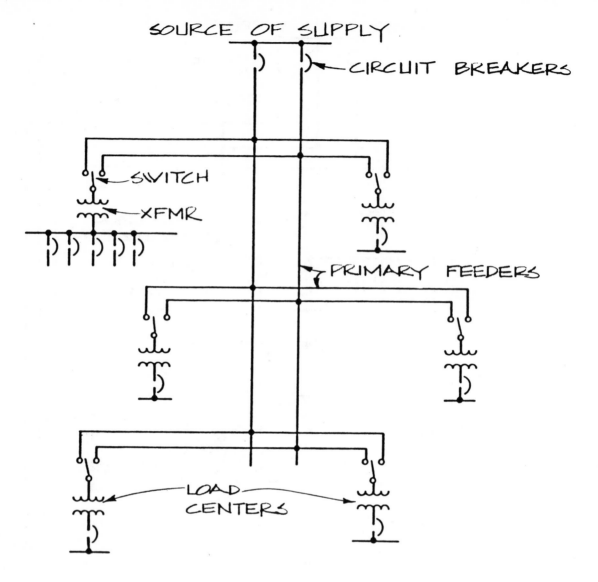

**FIGURE 3-71**

PRIMARY SELECTIVE-RADIAL
DISTRIBUTION SYSTEM

NO SCALE

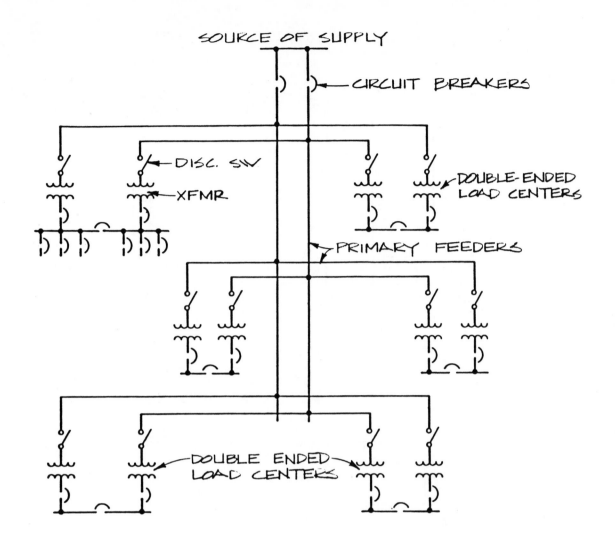

SOURCE OF SUPPLY

CIRCUIT BREAKERS

DISC. SW

XFMR

DOUBLE-ENDED LOAD CENTERS

PRIMARY FEEDERS

DOUBLE ENDED LOAD CENTERS

**FIGURE 3-72**

SECONDARY SELECTIVE-RADIAL
DISTRIBUTION SYSTEM

NO SCALE

317

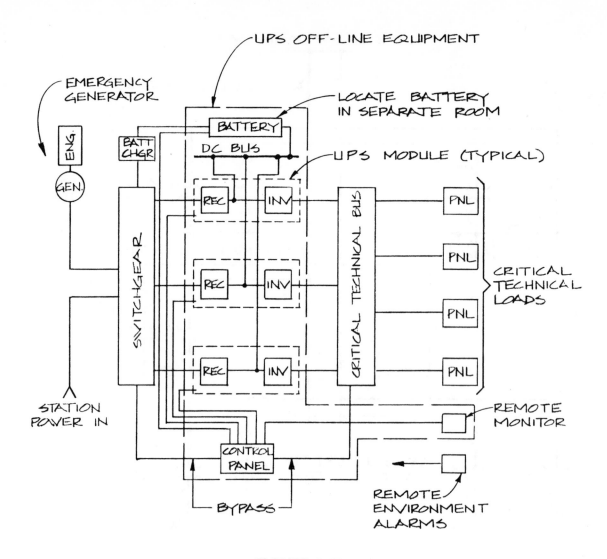

**FIGURE 3-73**

BLOCK DIAGRAM
UNINTERRUPTABLE POWER SYSTEM INSTALLATION
NO SCALE

Figure 3-73 is a schematic drawing indicating the basic block power
diagram of an uninterruptable power system supplied by an emergency
generator. Normally, the emergency generator is diesel fuel or perhaps
gasoline or liquid petroleum gas. The equipment must be carefully
located in a room that is properly ventilated. Arrangements and ventilation and code requirements (National and other) must be carefully
included.

318

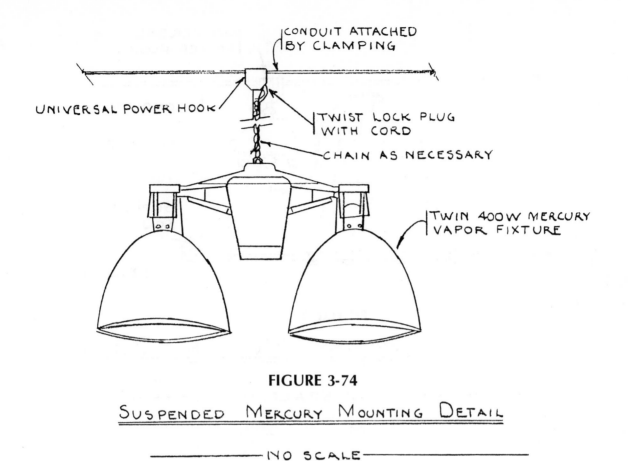

CONDUIT ATTACHED BY CLAMPING

UNIVERSAL POWER HOOK

TWIST LOCK PLUG WITH CORD

CHAIN AS NECESSARY

TWIN 400W MERCURY VAPOR FIXTURE

**FIGURE 3-74**

SUSPENDED MERCURY MOUNTING DETAIL

————————— NO SCALE —————————

Figures 3-74 and 3-75 are two of innumerable interior lighting diagrams that may be required. Obviously, if all lighting diagrams were included, the entire book could be filled with this type of work. These two are shown as an illustration of what is generally required in a lighting fixture delineation.

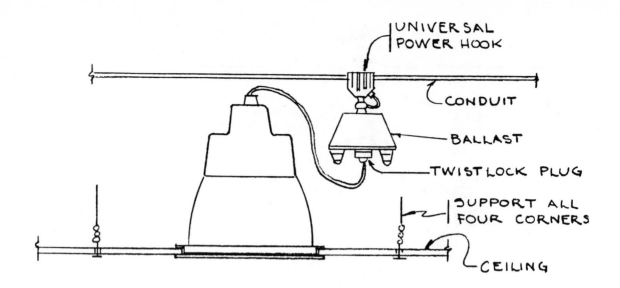

UNIVERSAL
POWER HOOK

CONDUIT

BALLAST

TWISTLOCK PLUG

SUPPORT ALL
FOUR CORNERS

CEILING

**FIGURE 3-75**

RECESSED MERCURY MOUNTING DETAIL

———— NO SCALE ————

320

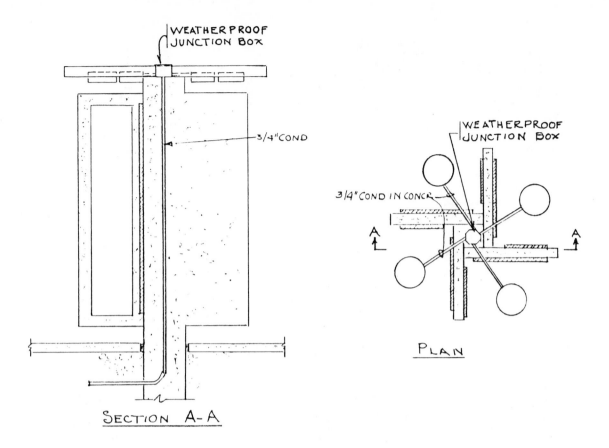

WEATHERPROOF
JUNCTION BOX

3/4"COND

SECTION A-A

WEATHERPROOF
JUNCTION BOX

3/4"COND IN CONCR

A

A

PLAN

**FIGURE 3-76**

KIOSK LIGHTING DETAIL

Figures 3-76, 3-77, and 3-78 are delineations of similar specialized fixtures used on the exterior. The three picked are relatively common in exterior lighting installations. There are, of course, many more that could have been included. Again, these are for guidance.

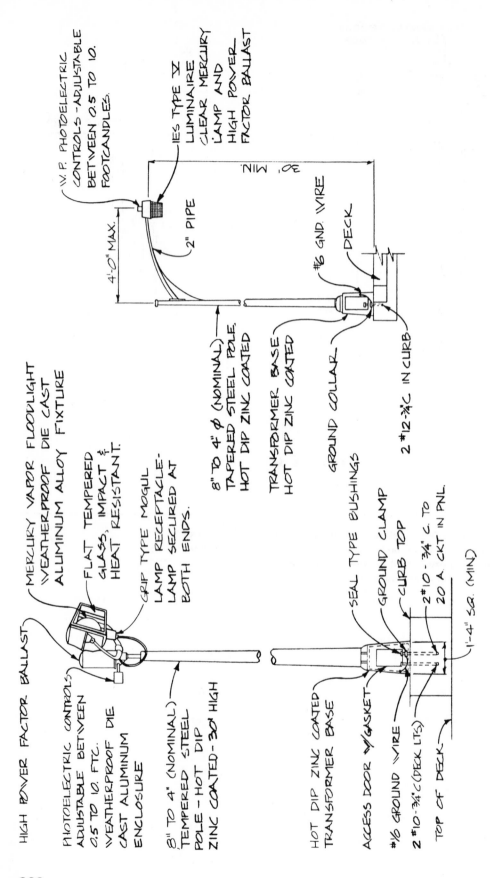

HIGH POWER FACTOR BALLAST

PHOTOELECTRIC CONTROLS ADJUSTABLE BETWEEN 0.5 TO 10 FTC. WEATHERPROOF DIE CAST ALUMINUM ENCLOSURE

8" TO 4" (NOMINAL) TEMPERED STEEL POLE - HOT DIP ZINC COATED - 30' HIGH

HOT DIP ZINC COATED TRANSFORMER BASE

ACCESS DOOR W/GASKET

#6 GROUND WIRE

2 #10-¾"C (DECK LTS)

TOP OF DECK

SEAL TYPE BUSHINGS

GROUND CLAMP

CURB TOP

2 #10 - ¾" C. TO 20 A. CKT IN PNL.

1'-4" SQ. (MIN)

W.P. PHOTOELECTRIC CONTROLS - ADJUSTABLE BETWEEN 0.5 TO 10. FOOTCANDLES.

IES TYPE Ⅴ LUMINAIRE CLEAR MERCURY LAMP AND HIGH POWER FACTOR BALLAST

4'-0" MAX.

2" PIPE

30' MIN.

#6 GND. WIRE

DECK

8" TO 4" Ø (NOMINAL) TAPERED STEEL POLE, HOT DIP ZINC COATED

TRANSFORMER BASE HOT DIP ZINC COATED

GROUND COLLAR

2 #12-¾"C IN CURB

MERCURY VAPOR FLOODLIGHT WEATHERPROOF DIE CAST ALUMINUM ALLOY FIXTURE

FLAT TEMPERED GLASS, IMPACT & HEAT RESISTANT.

GRIP TYPE MOGUL LAMP RECEPTACLE - LAMP SECURED AT BOTH ENDS.

**FIGURE 3-77**

LIGHTING STANDARD DETAILS

NO SCALE

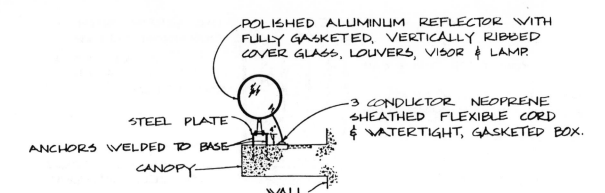

POLISHED ALUMINUM REFLECTOR WITH
FULLY GASKETED, VERTICALLY RIBBED
COVER GLASS, LOUVERS, VISOR & LAMP.

3 CONDUCTOR NEOPRENE
SHEATHED FLEXIBLE CORD
& WATERTIGHT, GASKETED BOX.

STEEL PLATE

ANCHORS WELDED TO BASE

CANOPY

WALL

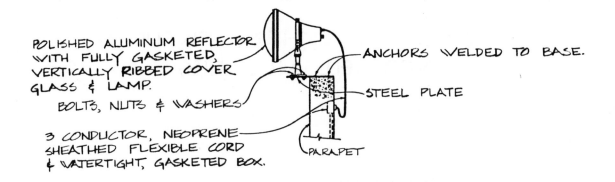

POLISHED ALUMINUM REFLECTOR
WITH FULLY GASKETED,
VERTICALLY RIBBED COVER
GLASS & LAMP.

BOLTS, NUTS & WASHERS

3 CONDUCTOR, NEOPRENE
SHEATHED FLEXIBLE CORD
& WATERTIGHT, GASKETED BOX.

ANCHORS WELDED TO BASE.

STEEL PLATE

PARAPET

## EXTERIOR FLOODLIGHTS

**FIGURE 3-78**

## PRISON LIGHTING DETAILS
### NO SCALE

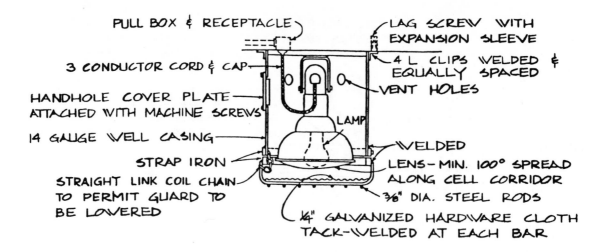

PULL BOX & RECEPTACLE

LAG SCREW WITH EXPANSION SLEEVE

3 CONDUCTOR CORD & CAP

4 L CLIPS WELDED & EQUALLY SPACED

HANDHOLE COVER PLATE ATTACHED WITH MACHINE SCREWS

VENT HOLES

LAMP

14 GAUGE WELL CASING

WELDED

STRAP IRON

LENS - MIN. 100° SPREAD ALONG CELL CORRIDOR

STRAIGHT LINK COIL CHAIN TO PERMIT GUARD TO BE LOWERED

3/8" DIA. STEEL RODS

1/4" GALVANIZED HARDWARE CLOTH TACK-WELDED AT EACH BAR

## EMERGENCY LIGHTING FIXTURES OVER CELL AIR CORES

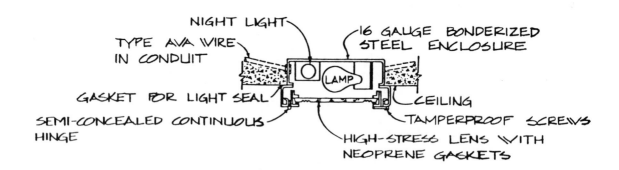

NIGHT LIGHT

TYPE AVA WIRE IN CONDUIT

16 GAUGE BONDERIZED STEEL ENCLOSURE

LAMP

GASKET FOR LIGHT SEAL

CEILING

SEMI-CONCEALED CONTINUOUS HINGE

TAMPERPROOF SCREWS

HIGH-STRESS LENS WITH NEOPRENE GASKETS

## UNIT INSTALLED OVER CELL GALLERIES AND EXERCISE CORRIDORS

**FIGURE 3-79**

## PRISON LIGHTING DETAILS

NO SCALE

Figures 3-79 and 3-80 illustrate the amount of detailing required for specialized fixtures. In this case, the fixtures shown are for a prison, but they could be for any number of other applications and are merely shown to indicate the amount of detailing that is required to properly show a light fixture.

**324**

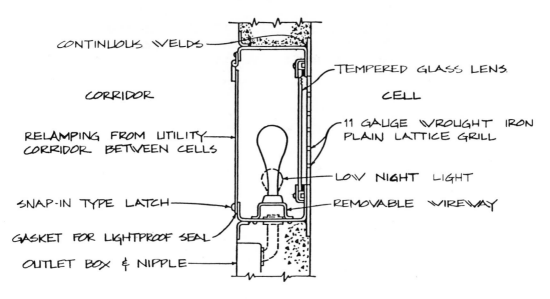

CONTINUOUS WELDS

CORRIDOR

RELAMPING FROM UTILITY
CORRIDOR BETWEEN CELLS

SNAP-IN TYPE LATCH

GASKET FOR LIGHTPROOF SEAL

OUTLET BOX & NIPPLE

TEMPERED GLASS LENS

CELL

11 GAUGE WROUGHT IRON
PLAIN LATTICE GRILL

LOW NIGHT LIGHT

REMOVABLE WIREWAY

LIGHTING FIXTURE
MAXIMUM SECURITY CELL

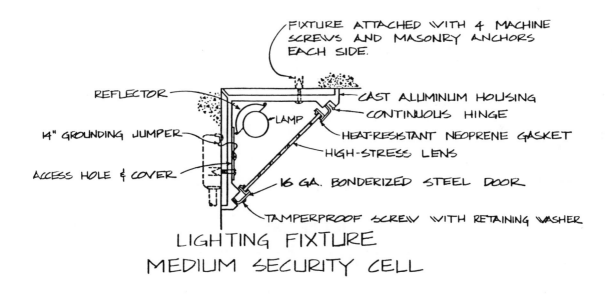

FIXTURE ATTACHED WITH 4 MACHINE
SCREWS AND MASONRY ANCHORS
EACH SIDE.

REFLECTOR

4" GROUNDING JUMPER

ACCESS HOLE & COVER

LAMP

CAST ALUMINUM HOUSING

CONTINUOUS HINGE

HEAT-RESISTANT NEOPRENE GASKET

HIGH-STRESS LENS

16 GA. BONDERIZED STEEL DOOR

TAMPERPROOF SCREW WITH RETAINING WASHER

LIGHTING FIXTURE
MEDIUM SECURITY CELL

**FIGURE 3-80**

PRISON LIGHTING DETAILS

NO SCALE

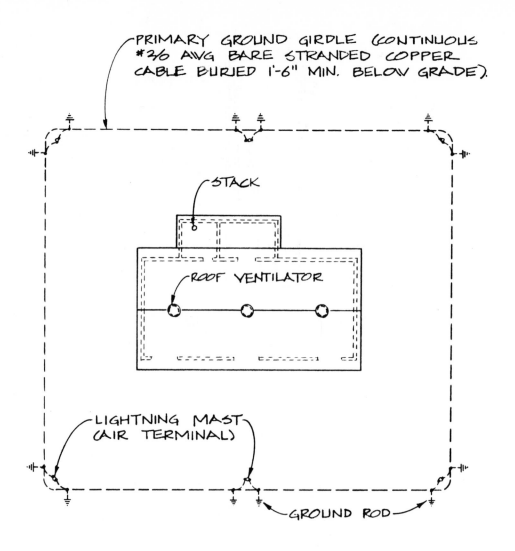

PRIMARY GROUND GIRDLE (CONTINUOUS #3/0 AWG BARE STRANDED COPPER CABLE BURIED 1'-6" MIN. BELOW GRADE).

STACK

ROOF VENTILATOR

LIGHTNING MAST (AIR TERMINAL)

GROUND ROD

PLAN

**FIGURE 3-81**

PRIMARY LIGHTNING PROTECTION SYSTEM

NO SCALE

Figures 3-81, 3-82, and 3-83 cover a relatively little-known and not often used subject of lightning protection. The details are not applicable per se to a given job, but they are presented in such a fashion that, with adjustment, could be incorporated into any work. They show, for example, the basic plan of a lightning protection system, the elevation, and notes that go with a typical lightning protection system, and the specialized details of grounding around the lightning mast. The details in that regard are self-explanatory.

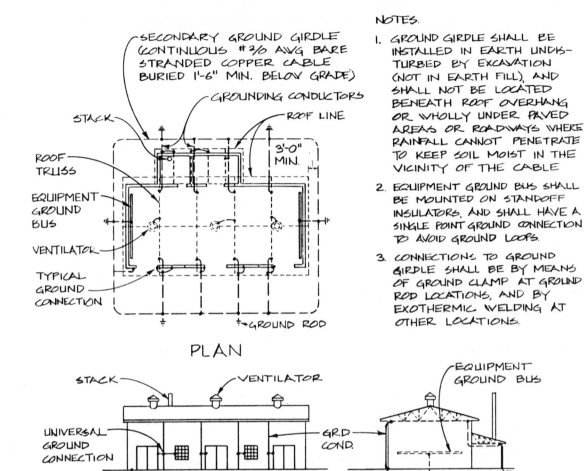

SECONDARY GROUND GIRDLE
(CONTINUOUS #3/0 AWG BARE
STRANDED COPPER CABLE
BURIED 1'-6" MIN. BELOW GRADE)

GROUNDING CONDUCTORS

STACK

ROOF LINE

ROOF TRUSS

3'-0" MIN.

EQUIPMENT GROUND BUS

VENTILATOR

TYPICAL GROUND CONNECTION

GROUND ROD

PLAN

NOTES.

1. GROUND GIRDLE SHALL BE INSTALLED IN EARTH UNDISTURBED BY EXCAVATION (NOT IN EARTH FILL), AND SHALL NOT BE LOCATED BENEATH ROOF OVERHANG OR WHOLLY UNDER PAVED AREAS OR ROADWAYS WHERE RAINFALL CANNOT PENETRATE TO KEEP SOIL MOIST IN THE VICINITY OF THE CABLE

2. EQUIPMENT GROUND BUS SHALL BE MOUNTED ON STANDOFF INSULATORS, AND SHALL HAVE A SINGLE POINT GROUND CONNECTION TO AVOID GROUND LOOPS.

3. CONNECTIONS TO GROUND GIRDLE SHALL BE BY MEANS OF GROUND CLAMP AT GROUND ROD LOCATIONS, AND BY EXOTHERMIC WELDING AT OTHER LOCATIONS.

STACK

VENTILATOR

EQUIPMENT GROUND BUS

UNIVERSAL GROUND CONNECTION

GRD. COND.

GROUND OUTLET FOR REINF. STL.

GROUND ROD

FRONT ELEV.

SIDE ELEV.

**FIGURE 3-82**

SECONDARY LIGHTNING PROTECTION SYSTEM

NO SCALE

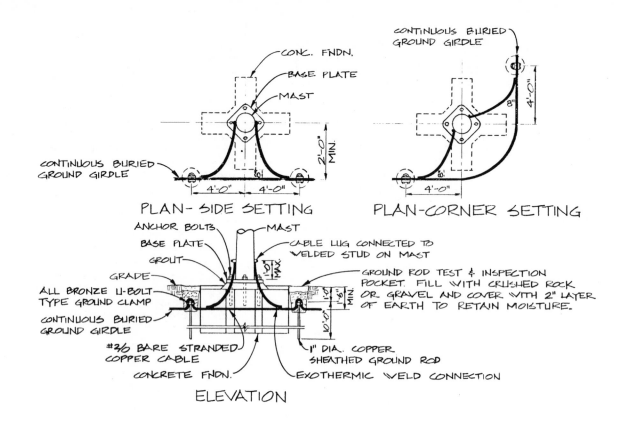

CONC. FNDN.
BASE PLATE
MAST
CONTINUOUS BURIED GROUND GIRDLE
2'-0" MIN.
4'-0"    4'-0"
6"

PLAN – SIDE SETTING

CONTINUOUS BURIED GROUND GIRDLE
4'-0"
8"
8"
4'-0"

PLAN-CORNER SETTING

ANCHOR BOLTS
MAST
BASE PLATE
CABLE LUG CONNECTED TO WELDED STUD ON MAST
GROUT
GRADE
1'-0" MAX.
GROUND ROD TEST & INSPECTION POCKET. FILL WITH CRUSHED ROCK OR GRAVEL AND COVER WITH 2" LAYER OF EARTH TO RETAIN MOISTURE.
ALL BRONZE U-BOLT TYPE GROUND CLAMP
CONTINUOUS BURIED GROUND GIRDLE
1'-6" MIN.
10'-0"
#3/0 BARE STRANDED COPPER CABLE
1" DIA. COPPER SHEATHED GROUND ROD
CONCRETE FNDN.
EXOTHERMIC WELD CONNECTION

ELEVATION

**FIGURE 3-83**

LIGHTNING MAST GROUND CONNECTION DETAILS

NO SCALE

328

PANEL SCHEDULE

| | | PANELS | | | BRANCH CIRCUITS | | | | | | | | REMARKS |
|-----|------|----------|-----|-------|------|-------|------|-------|--------|-------|--------|-------|---------|
| KEY | TYPE | LOCATION | MTG | MAINS | TYPE | FRAME | TRIP | POLES | ACTIVE | SPARE | TOTAL | | |
| | | | | | | | | | | | | | |
| | | | | | | | | | | | | | |
| | | | | | | | | | | | | | |
| | | | | | | | | | | | | | |

**FIGURE 3-84**

Figures 3-84 and 3-85 are two commonly used schedules. Figure 3-84 shows a panel schedule that could be enlarged or slightly altered to become a panel and distribution schedule. Figure 3-85 is a lighting fixture schedule. It is not usual to specify lighting fixtures in the specifications, but instead they are listed on the plan. The schedules for motors, etc., are shown in the mechanical schedule section. Some of these schedules could be incorporated in the electrical, if so desired.

329

| LIGHTING FIXTURE SCHEDULE | | | | | | |
|---|---|---|---|---|---|---|
| KEY | MFR. | CAT. NO. | DESCRIPTION | LAMP | | REMARKS |
| | | | | QUAN. | TYPE | |
| | | | | | | |

**FIGURE 3-85**

Crouse-Hinds
Code Digest

Appendix III

Installation
Diagram
for Sealing

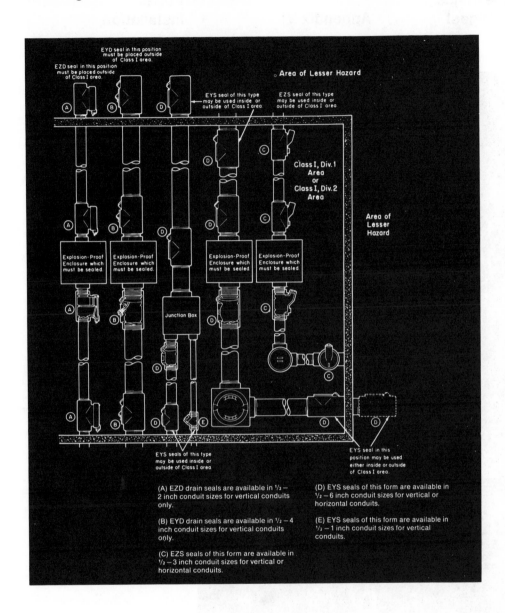

FIGURE 3-86

Figures 3-86, 3-87, and 3-88 are typical examples of explosion-proof work
and are reprinted directly with the express approval of the Crouse-Hinds
Company, of Syracuse, NY. These particular details are extremely well
thought out, and the manufacturer has many more that can be used and
can be secured upon request. The importance of these details is that they
illustrate, for instance, the sealing which is commonly overlooked on a
drawing, the drains—the actual, physical size in relative proportion to a
starter, and the type of system used for class 1 versus class 2 work.

**Crouse-Hinds
Code Digest** **Appendix VI**

**Diagram
for Class I, Div. 1
Power
Installation**

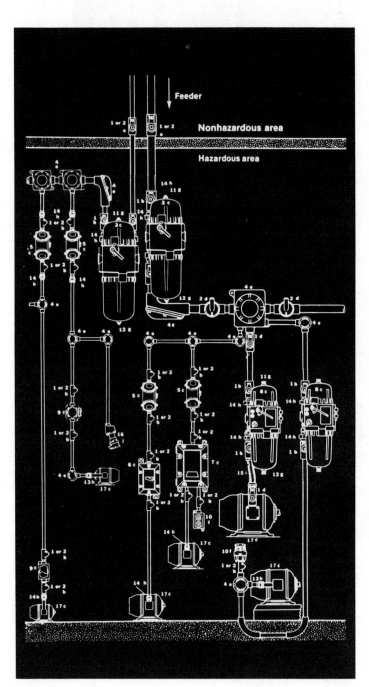

### Key to Numerals

**1** Sealing fitting. EYS for horizontal or vertical.

**2** Sealing fitting. EZS for vertical or horizontal conduits.

**3** Circuit breaker EPC.

**4** Junction box Series GUA, GUB and GUJ have threaded covers. Series CPS and Type LBH have ground flat surface covers.

**5** Circuit breaker FLB.

**6** Manual line starter EMN.

**7** Magnetic line starter EMG.

**8** Combination circuit breaker and line starter EPC.

**9** Switch or motor starter. Series EFS, EDS, or GUSC.

**10** Push button station. Series EFS or OFC.

**11** Breather. ECD.

**12** Drain. ECD.

**13** Union. UNF.

**14** Union. UNY.

**15** Flexible coupling. ECH.

**16** Plug receptacle. CES. Factory sealed.

**17** Motor for hazardous location.

### National Electrical Code References

**a** Sec. 501-5a-4. Seals required where conduits pass from hazardous to non-hazardous area.

**b** Sec. 501-5a-1. Seals required within 18 inches of all arcing devices.

**c** Art. 430 should be studied for detailed requirements for conductors, motor feeders, motor feeder and motor branch circuit protection, motor overcurrent protection, motor controllers and motor disconnecting means.

**d** Sec. 501-5a-2. Seals required if conduit is 2 inches or larger in size.

**e** Sec. 501-4a. All boxes must be explosion-proof and threaded for rigid or IMC conduit.

**f** Sec. 501-6a. Push button stations must be explosion-proof.

**g** Sec. 501-5f. Breathers and drains needed in all humid locations.

**h** Sec. 501-4a. All joints and fittings must be explosion-proof.

**i** Sec. 501-4a. Flexible connections must be explosion-proof.

**j** Sec. 501-12. Receptacles and plugs must be explosion-proof, and provide grounding connections for portable devices.

**FIGURE 3-87**

**Appendix VIII**

Diagram
for Class II
Power
Installation

Crouse-Hinds
Code Digest

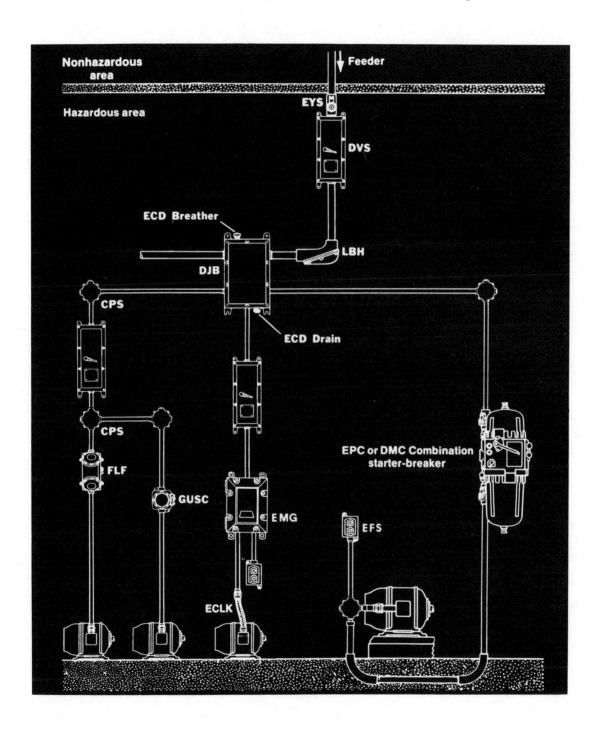

**FIGURE 3-88**

# Index